T0401990

27

Topics in Organometallic Chemistry

Topics in Organometallic Chemistry
Recently Published and Forthcoming Volumes

Molecular Organometallic Materials for Optics
Volume Editors: H. Le Bozec, V. Guerchais
Vol. 28, 2009

Conducting and Magnetic Organometallic Molecular Materials
Volume Editors: M. Fourmigué, L. Ouahab
Vol. 27, 2009

Metal Catalysts in Olefin Polymerization
Volume Editor: Z. Guan
Vol. 26, 2009

Bio-inspired Catalyst
Volume Editor: T. R. Ward
Vol. 25, 2009

Directed Metallation
Volume Editor: N. Chatani
Vol. 24, 2007

Regulated Systems for Multiphase Catalysis
Volume Editors: W. Leitner, M. Hölscher
Vol. 23, 2008

Organometallic Oxidation Catalysis
Volume Editors: F. Meyer, C. Limberg
Vol. 22, 2007

N-Heterocyclic Carbenes in Transition Metal Catalysis
Volume Editor: F. Glorius
Vol. 21, 2006

Dendrimer Catalysis
Volume Editor: L. H. Gade
Vol. 20, 2006

Metal Catalyzed Cascade Reactions
Volume Editor: T. J. J. Müller
Vol. 19, 2006

Catalytic Carbonylation Reactions
Volume Editor: M. Beller
Vol. 18, 2006

Bioorganometallic Chemistry
Volume Editor: G. Simonneaux
Vol. 17, 2006

Surface and Interfacial Organometallic Chemistry and Catalysis
Volume Editors: C. Copéret, B. Chaudret
Vol. 16, 2005

Chiral Diazaligands for Asymmetric Synthesis
Volume Editors: M. Lemaire, P. Mangeney
Vol. 15, 2005

Palladium in Organic Synthesis
Volume Editor: J. Tsuji
Vol. 14, 2005

Metal Carbenes in Organic Synthesis
Volume Editor: K. H. Dötz
Vol. 13, 2004

Theoretical Aspects of TransitionMetal Catalysis
Volume Editor: G. Frenking
Vol. 12, 2005

Ruthenium Catalysts and Fine Chemistry
Volume Editors: C. Bruneau, P. H. Dixneuf
Vol. 11, 2004

New Aspects of Zirconium Containing
Organic Compounds
Volume Editor: I. Marek
Vol. 10, 2004

Precursor Chemistry of Advanced Materials
CVD, ALD and Nanoparticles
Volume Editor: R. Fischer
Vol. 9, 2005

Metallocenes in Stereoselective Synthesis
Volume Editor: T. Takahashi
Vol. 8, 2004

Conducting and Magnetic Organometallic Molecular Materials

Volume Editors: Marc Fourmigué and Lahcène Ouahab

With Contributions by

Manuel Almeida · Olivier Cador · Toshiaki Enoki · Christophe Faulmann · Marc Fourmigué · Vasco Gama · Stéphane Golhen · Reizo Kato · Kazuya Kubo · Akira Miyazaki · Lahcène Ouahab · John A. Schlueter · Lydie Valade

Editors

Dr. Marc Fourmigué
Sciences Chimiques de Rennes
Equipe Matière Condensée et
Systèmes Electroactifs (MaCSE)
UMR 6226 CNRS-Université de Rennes 1
Campus de Beaulieu
35042 Rennes Cedex, France
marc.fourmigue@univ-rennes1.fr

Dr. Lahcène Ouahab
Laboratoire Sciences Chimiques de Rennes
Equipe Organometalliques et Matériaux
Moléculaires
UMR 6226 CNRS-Université de Rennes 1
Campus de Beaulieu, Bat 10B, bureau 105
263, Avenue du Général Leclerc
35042 Rennes Cedex, France
lahcene.ouahab@univ-rennes1.fr

ISSN 1436-6002 e-ISSN 1616-8534
ISBN 978-3-642-00407-0 e-ISBN 978-3-642-00408-7
DOI: 10.1007/978-3-642-00408-7
Springer Dordrecht Heidelberg London New York

Library of Congress Control Number: 2009922221

Cover design: KünkelLopka GmbH; *volume cover:* SPi Publisher Services

Printed on acid-free paper

Springer is part of Springer Science+Business Media (www.springer.com)

Volume Editors

Dr. Marc Fourmigué
Sciences Chimiques de Rennes
Equipe Matière Condensée et
Systèmes Electroactifs (MaCSE)
UMR 6226 CNRS-Université de Rennes 1
Campus de Beaulieu
35042 Rennes Cedex, France
marc.fourmigue@univ-rennes1.fr

Dr. Lahcène Ouahab
Laboratoire Sciences Chimiques de Rennes
Equipe Organometalliques et Matériaux
Moléculaires
UMR 6226 CNRS-Université de Rennes 1
Campus de Beaulieu, Bat 10B, bureau 105
263, Avenue du Général Leclerc
35042 Rennes Cedex, France
lahcene.ouahab@univ-rennes1.fr

Editorial Board

Topics in Organometallic Chemistry Also Available Electronically

Topics in Organometallic Chemistry is included in Springer's eBook package *Chemistry and Materials Science*. If a library does not opt for the whole package the book series may be bought on a subscription basis. Also, all back volumes are available electronically.

For all customers who have a standing order to the print version of *Topics in Organometallic Chemistry*, we offer the electronic version via SpringerLink free of charge.

If you do not have access, you can still view the table of contents of each volume and the abstract of each article by going to the SpringerLink homepage, clicking on "Chemistry and Materials Science," under Subject Collection, then "Book Series," under Content Type and finally by selecting *Topics in Organometallic Chemistry*.

You will find information about the

- Editorial Board
- Aims and Scope
- Instructions for Authors
- Sample Contribution

at springer.com using the search function by typing in *Topics in Organometallic Chemistry*.
Color figures are published in full color in the electronic version on SpringerLink.

Aims and Scope

The series *Topics in Organometallic Chemistry* presents critical overviews of research results in organometallic chemistry. As our understanding of organometallic structures, properties and mechanisms grows, new paths are opened for the design of organometallic compounds and reactions tailored to the needs of such diverse areas as organic synthesis, medical research, biology and materials science. Thus the scope of coverage includes a broad range of topics of pure and applied organometallic chemistry, where new breakthroughs are being made that are of relevance to a larger scientific audience.

The individual volumes of *Topics in Organometallic Chemistry* are thematic. Review articles are generally invited by the volume editors.

In references *Topics in Organometallic Chemistry* is abbreviated Top Organomet Chem and is cited as a journal.

Preface

For several years, the two parallel worlds of Molecular Conductors in one hand and Molecular Magnetism in the other have grown side by side, the former essentially based on radical organic molecules, the latter essentially based on the high spin properties of metal complexes. Over the last few years however, *organometallic* derivatives have started to play an increasingly important role in both worlds, and have in many ways contributed to open several passages between these two worlds. This volume recognizes this important emerging evolution of both research areas. It is not intended to give a comprehensive view of all possible organometallic materials, and polymers for example were not considered here. Rather we present a selection of the most recent research topics where organometallic derivatives were shown to play a crucial role in the setting of conducting and/or magnetic properties in crystalline materials. First, the role of organometallic anions in tetrathiafulvalenium-based molecular conductors is highlighted by Schlueter, while Kubo and Kato describe very recent *ortho*-metalated chelating ligands appended to the TTF core and their conducting salts. The combination of conducting and magnetic properties and the search for π–d interactions are analyzed in two complementary contributions by Myazaki and Ouahab, while Valade focuses on the only class of metal bis(dithiolene) complexes to give rise to superconductive molecular materials, in association with organic as well as organometallic cations. The structures and properties of the salts based on such metallocenium cations and transition metal bis-dichalcogenide anions are then comprehensively reviewed by Almeida and Gama. This is followed by a review by Fourmigué on paramagnetic organometallic cyclopentadienyl/dichalcogenide complexes.
Perhaps the common characteristic of all contributions to this volume is the permanent concern about the intimate relationships between the structural and electronic properties. Indeed, the careful design of increasingly complex molecular and supramolecular architectures allows us now to anticipate many molecular and solid state properties, but the final solid state structures are always the results of many competing interactions. The resulting electronic properties of these radical assemblies, whether conductivity or magnetism, are always very sensitive to minute modifications of their solid state structures and one of the main difficulties through

the investigation of such materials is always the identification of the nature and relative strength of all possible electronic interactions paths. In that respect, organometallic derivatives, for example with strongly delocalized spin densities, bring added elements of complexity, as the reader will discover in the next seven contributions.

Rennes, France,
December 2008

Marc Fourmigué
Lahcène Ouahab

Contents

Contents to
Topics in Organometallic Chemistry Volume 28
Molecular Organometallic Materials for Optics

Volume Editors: Hubert Le Bozec and Véronique Guerchais

Top Organomet Chem (2009) 27: 1–33

Tetrathiafulvalene-Based Conductors Containing Organometallic Components

John A. Schlueter

Abstract Hundreds of cation radical salts containing tetrathiafulvalene (TTF) and its derivatives have been prepared and characterized over the past 50 years. After superconductivity was discovered in salts containing bis(ethylenedithio)tetrathiafulvalene (BEDT-TTF), a tremendous international effort was undertaken to prepare cation radical salts containing nearly every imaginable type of anion in a search for new superconducting materials. Surprisingly, only a relatively small number of these salts contained organometallic anions. The use of tetrakis(trifluoromethyl)metallate anions as charge compensating components in BEDT-TTF salts resulted in the discovery of a diverse family of molecular superconductors. More recently, organometallic anions have been utilized as a means to introduce multiple properties into a single material. Halophenylene, metallocarborane and metallocene-type anions have been utilized as components of cation radical salts. Several covalent derivatives of the TTF molecule have also been reported that incorporate organometallic moieties. Due to the diverse nature of organometallic anions and contemporary interest in coupling redox active species to the TTF moiety, it is likely that this field will exhibit significant growth in the years ahead.

Keywords Cation-radical salts, Molecular conductors, Organic superconductors, Organometallic anions, Tetrakis(trifluoromethyl)metallates, Tetrathiafulvalene

Contents

J.A. Schlueter
Materials Science Division, Argonne National Laboratory, Argonne, IL, 9700 S. Cass Avenue, Argonne, IL 60439, USA, E-mail: JASchlueter@anl.gov

M. Fourmigué and L. Ouahab (eds.), *Conducting and Magnetic Organometallic Molecular Materials*, Topics in Organometallic Chemistry 27,
DOI: 10.1007/978-3-642-00408-7_1, © Springer-Verlag Berlin Heidelberg 2009

Abbreviations

BEDT-TTF	Bis(ethylenedithio)tetrathiafulvalene
BETS	Bis(ethylenedithio)tetraselenafulvalene
Cp	Cyclopentyldienyl
Cp*	Pentamethylcyclopentyldienyl
DBCE	1,2-Dibromo-1-chloroethane
Dca	Dicyanamide
DDQ	2,3-Dichloro-5,6-dicyanobenzoquinone
$DmitH_2$	4,5-Dimercapto-1,3-dithione-2-thione
EDAX	Energy dispersive analysis by X-ray
EPR	Electron paramagnetic resonance
o-Me_2TTF	3,4-Dimethyltetrathiafulvalene
MI	Metal-to-insulator
PPN	Bis(triphenylphosphoranylidene)ammonium
Ppy	*C*-dehydro-2-phenylpyridine
TBA	Tetrabutylammonium
TCE	1,1,2-Trichloroethane
TCNQ	Tetracyanoquniodimethane
TEA	Tetraethylammonium
TMA	Tetramethylammonium
TMTSF	Bis(tetramethyl)tetraselenafulvalene
TMTTF	Bis(tetramethyl)tetrathiafulvalene
TTF	Tetrathiafulvalene
TTFFc	1-Tetrathiafulvalenylferrocene
$TTFPh_2$	4,4′-Diphenyltetrathiafulvalene
VT	Bis(vinylenedithio)tetrathiafulvalene

1 Introduction

The synthesis of the bis-1,3-dithiolium radical cation (TTF^{+}) in 1970 [1] enabled dramatic growth in the field of molecular conductors in the decades thereafter. TTF and several of its homologues are depicted in Scheme 1. The field of low dimensional molecular metals was further motivated by the discovery of the TTF-TCNQ charge-transfer complex in 1973 [2, 3]. Seven years later, superconductivity was induced in the cation-radical salt, $(TMTSF)_2PF_6$, upon application of 12 kbar pressure [4]. Shortly thereafter, superconductivity below 1.4 K was observed at ambient pressure in the perchlorate analog [5].

The electron-donor molecule most studied as a component of cation-radical salts is BEDT-TTF. Over 50 superconducting salts containing this molecule have been reported in the literature. Several distinct packing motifs [6–8] of the BEDT-TTF radical cations have yielded superconducting salts, but the most studied of these are

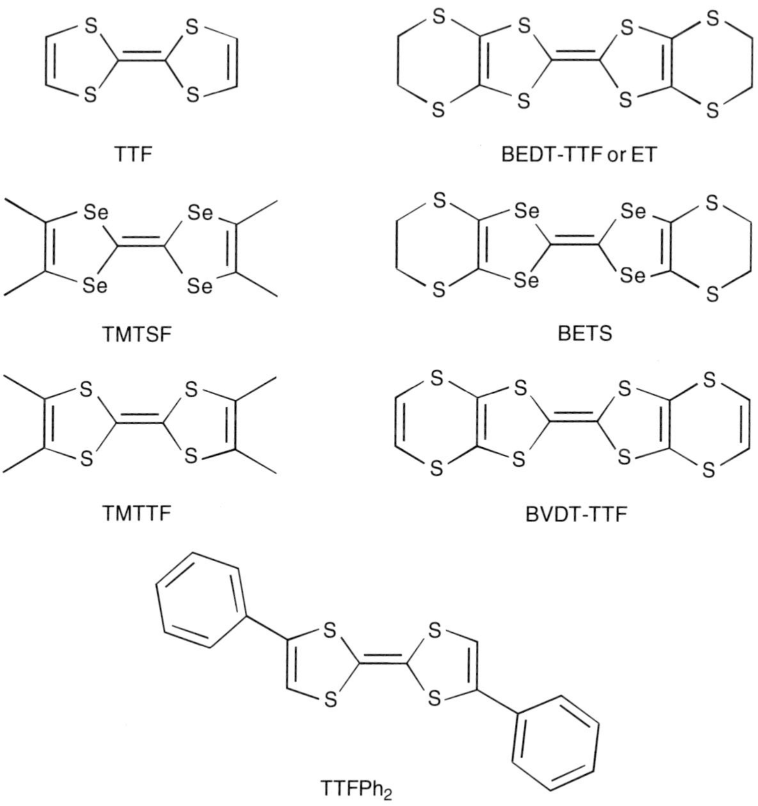

TTF

BEDT-TTF or ET

TMTSF

BETS

TMTTF

BVDT-TTF

$TTFPh_2$

Scheme 1 Electron-donor molecules that have been utilized as components of cation radical salts with organometallic anions

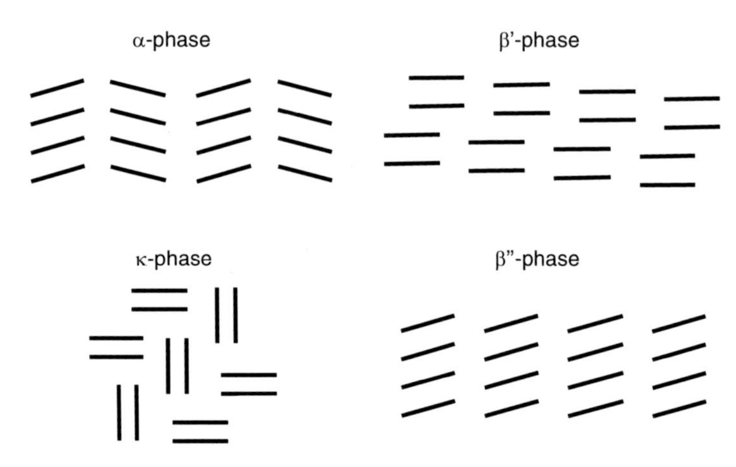

Scheme 2 Four common packing motifs of the BEDT-TTF molecule. The *black bars* represent and end-on view of the molecule, looking down the long molecular axis. The α-phase, which is quite similar to the θ-phase, is characterized as a herringbone structure, the β′-phase contains stacks of slipped dimers, the κ-phase contains orthogonally arranged dimers, and the β″-phase is characterized by canted stacks of donor molecules

the κ-type which is characterized by orthogonally arranged dimers. A few of the more common packing motifs are illustrated in Scheme 2. The κ-(BEDT-TTF)$_2$Cu(NCS)$_2$ salt has a superconducting transition temperature, T_c, of 10.4 K [9]. Among cation radical salts, κ-(BEDT-TTF)$_2$Cu[N(CN)$_2$]X (X = Br and Cl) have the highest T_cs: κ-(BEDT-TTF)$_2$Cu[N(CN)$_2$]Br (T_c =11.6 K [10]) and κ-(BEDT-TTF)$_2$Cu[N(CN)$_2$]Cl (T_c =12.8 K at 300 bar [11]).

A wide range of anions have been investigated as components of cation radical salts including halides, trihalides, polyhalides, metal halides of various geometries, polynuclear clusters, etc. Classes of anions that have been studied to a lesser extent include organic and organometallic anions [12]. The organometallic anions that have been studied to date are depicted in Scheme 3. Several excellent reviews have been written that cover various aspects of this growing field [13–21]. This review focuses on progress that has been made through the use of tetrathiafulvalene-based materials that incorporate organometallic components that contain metal-carbon bonds. Several such classes of these materials have been reported, including cation radical salts, neutral coordination complexes and cocrystals.

2 Tetrakis(trifluoromethyl)metallates, $M(CF_3)_4^-$

Among organometallic anions, those of the $M(CF_3)_4^-$ (M = Cu, Ag and Au) family have the been most studied as components of cation radical salts. These anions were originally chosen because of their surprising stability and solubility in organic

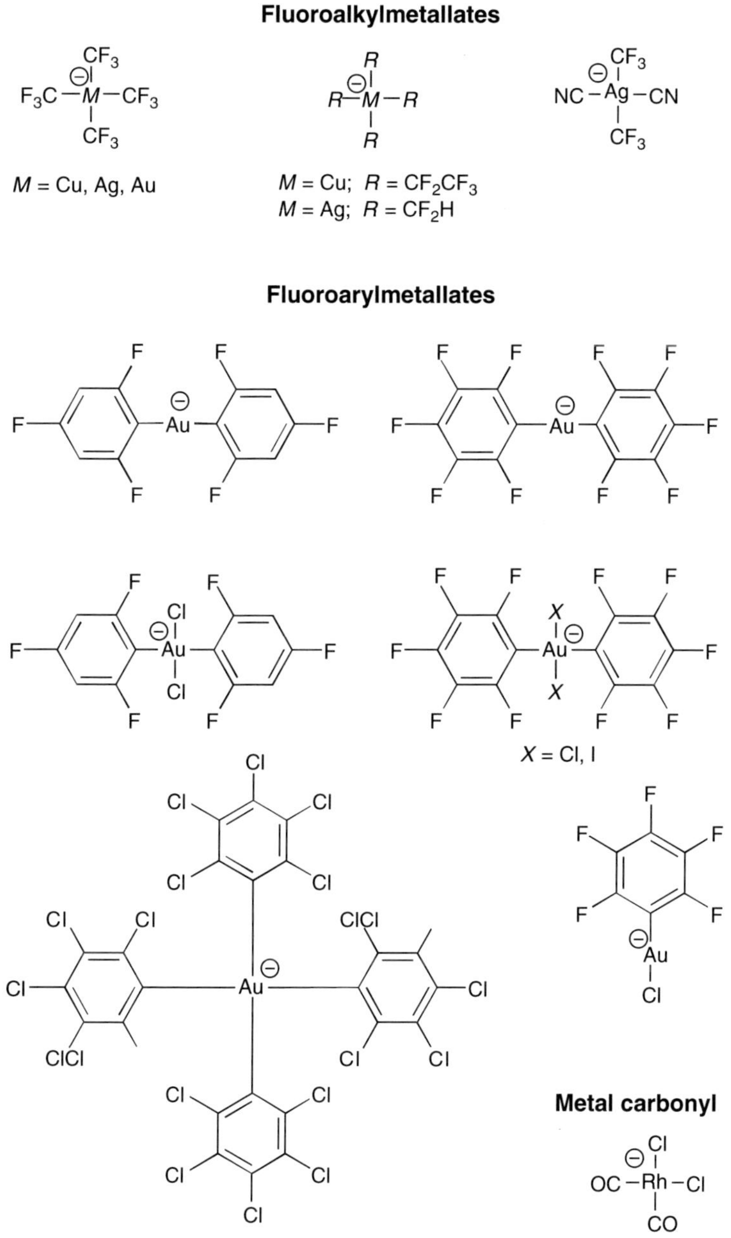

Fluoroalkylmetallates

M = Cu, Ag, Au

M = Cu; $R = CF_2CF_3$
M = Ag; $R = CF_2H$

Fluoroarylmetallates

X = Cl, I

Metal carbonyl

Scheme 3 (Continued)

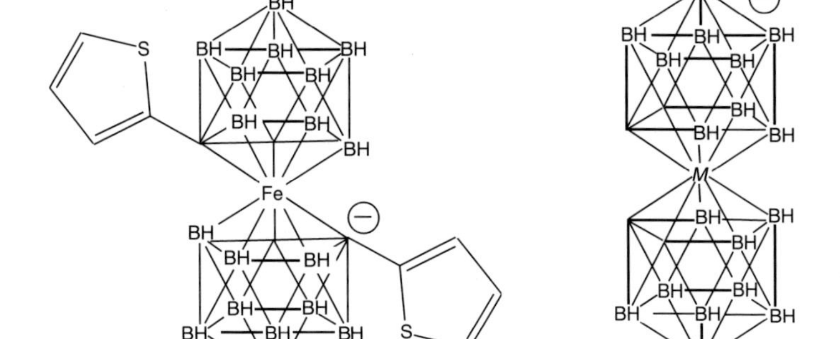

Scheme 3 Organometallic anions that have been used as components of cation radical salts

solvents. The molecular structures of these anions are quite similar and are characterized by an approximately square-planar arrangement of the four carbon atoms around the metal nucleus [22, 23]. The 12 electronegative fluorine atoms that shield the metal center are also useful for forming important hydrogen bonds with the ethylene groups of the BEDT-TTF molecules. Among these BEDT-TTF salts, four distinct phases have been reported that can be identified based on their morphology and EPR line widths. Three of these phases contain cocrystallized solvent molecules. Superconducting ground states have been observed in two of these phases. The κ_H-phase has superconducting transition temperatures that approach the record for cation-radical salts.

2.1 κ_L-(BEDT-TTF)$_2$M(CF$_3$)$_4$(1,1,2-trihaloethane)

As illustrated in Fig. 1, this phase can be visually identified by its hexagonal plate-like geometry. A full crystal structure has been reported for the κ_L-(BEDT-TTF)$_2$M (CF$_3$)$_4$(TCE) (M = Cu and Ag) members of this family [24] which contains layers of BEDT-TTF radical cations separated by layers containing $[M(CF_3)_4]^-$ anions and TCE solvent molecules (see Fig. 2). The BEDT-TTF molecules pack in a typical κ-type motif [7] that is characterized by orthogonally arranged dimers. The anionic layers are the thickest known (~8.3 Å) for molecular superconductors. This is a result of the surprising fact that the anions are arranged with their molecular plane perpendicular to the BEDT-TTF layers. It is interesting to note that at 115 K, both the anions and solvent molecules are crystallographically disordered. It is known that such disorder can be detrimental to superconductivity [25] and structural studies are in progress to determine whether an anion ordering occurs at lower temperatures.

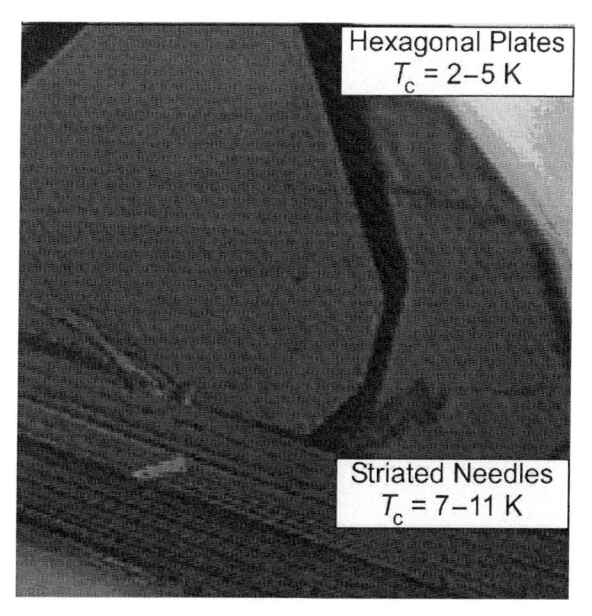

Fig. 1 Photograph illustrating the morphologies characteristic of the κ_L- and κ_H-phases of (BEDT-TTF)$_2$M(CF$_3$)$_4$(solvent)

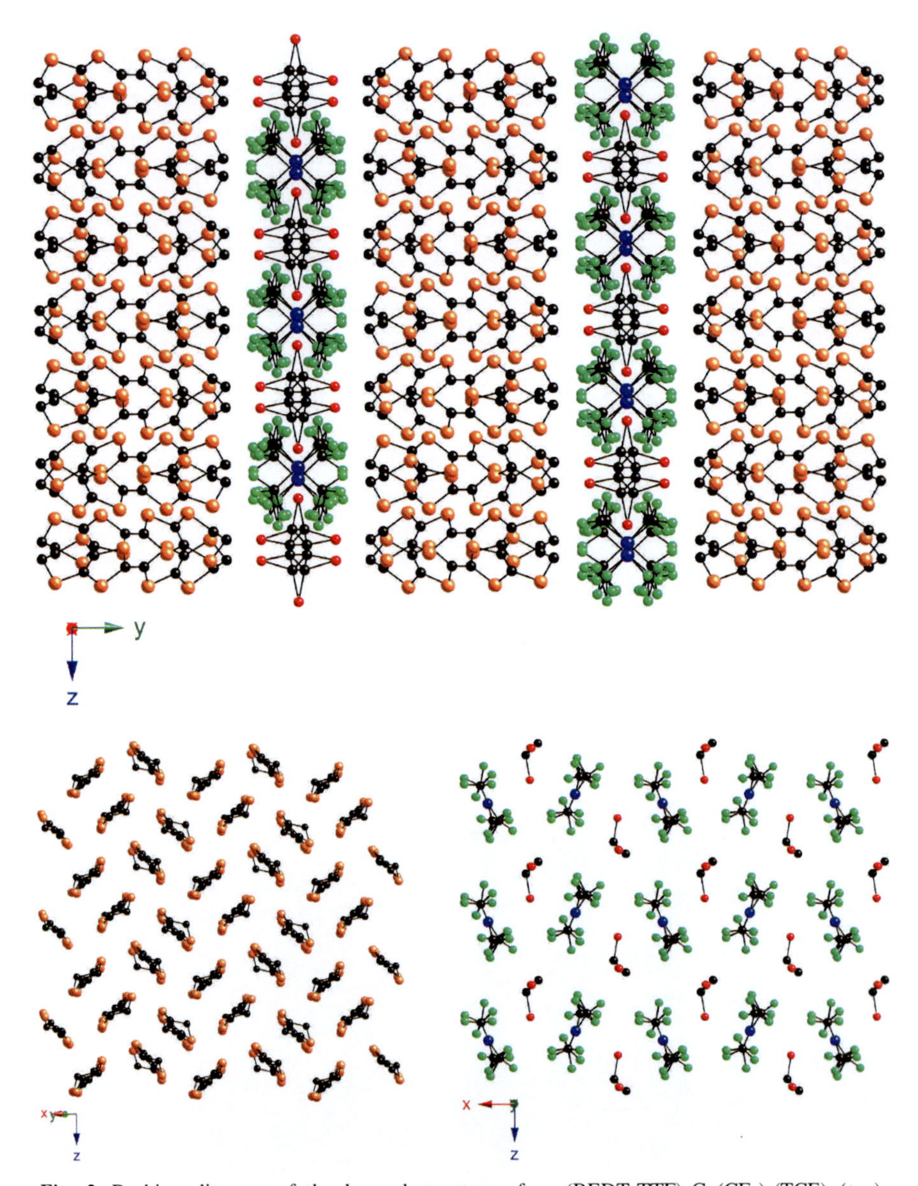

Fig. 2 Packing diagram of the layered structure of κ_L-(BEDT-TTF)$_2$Cu(CF$_3$)$_4$(TCE) (*top*). The packing motif of the BEDT-TTF electron donor molecules (*lower left*). The anion layer contains both disordered [Cu(CF$_3$)$_4$]$^-$ anions and neutral TCE solvent molecules (*lower right*). In all cases, hydrogen atoms have been omitted for clarity

Superconductivity was initially discovered in the κ_L-(BEDT-TTF)$_2$M(CF$_3$)$_4$ (TCE) members of this family with T_c = 4.0 K (M = Cu) [26], 2.6 K (Ag) [27] and 2.1 K (Au) [28]. In addition to exchange of the metal atom, it was found that new superconducting salts could be formed by replacing the chorine atoms on the TCE solvent molecule with bromine. Through use of the 2-bromo-1,1-dichloroethane, 1-bromo-1,2-dichloroethane, 1,2-dibromo-1-chloroethane, and 1,1,2-tribromoethane solvent, 12 additional superconducting salts were prepared. As illustrated in Fig. 3, the T_cs for this family range between 2.1 K and 5.8 K [29]. It is expected that additional superconducting salts would be formed through use of the 1,1-dibromo-2-chloroethane solvent, but this chemical is synthetically challenging to prepare. Attempts have been made to incorporate a wide variety of solvent molecules in these salts, but only the 1,1,2-trihaloethanes in which the halogen is chlorine or bromine form solvated salts [30]. It is believed that fluorinated ethanes compete too strongly for hydrogen bonding sites located on the BEDT-TTF ethylene groups, thus preempting formation of an isostructural phase.

An initial attempt was made to correlate the structural properties of the κ_L-(BEDT-TTF)$_2$M(CF$_3$)$_4$(TCE) salts with their superconducting transition temperature [31]. The relationship between T_c and any single unit cell parameter failed to show any discernable trend. The best correlation was obtained by plotting T_c as a function of the b/c ratio, where b is the interlayer and c is an intralayer direction. A similar conclusion was reached through the determination of uniaxial pressure coefficients of β"-(BEDT-TTF)$_2$SF$_5$CH$_2$CF$_2$SO$_3$ and κ-(BEDT-TTF)$_2$Cu(NCS)$_2$ through the measurement of thermal expansion [32]. These results also indicated that expansion of the interlayer direction and compression of an intralayer direction

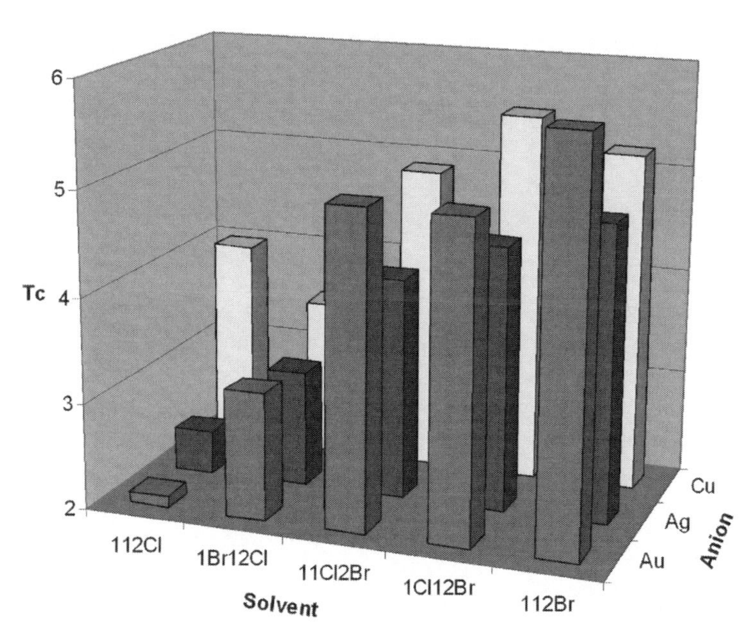

Fig. 3 Chart illustrating the superconducting transition temperatures as a function of solvent and anion in the κ_L-(BEDT-TTF)M(CF$_3$)$_4$(1,1,2-trihaloethane) (M = Cu, Ag and Au) series. 112Cl is 1,1,2-trichoroethane, 1Br12Cl is 1-bromo-1,2-dichloroethane, 11Cl2Br is 1,1-dichloro-2-bromoethane, 1Cl12Br is 1-chloro-1,2-dibromoethane and 112Br is 1,1,2-tribromoethane

lead to increased T_cs. This conclusion is further supported by a quantitative comparison of the T_cs of the four κ_L-(BEDT-TTF)$_2$Ag(CF$_3$)$_4$(1,1,2-trihaloethane) salts where the number of bromine atoms on the solvent molecule is incrementally increased from zero to three: the T_cs increase with the addition of the larger bromine atoms which force the conducting layers further apart [33]. Further evidence for this conclusion lies in a comparison of the T_cs of the six 1-bromo-1,2-dichloroethane and 2-bromo-1,1-dichloroethane salts of κ_L-(BEDT-TTF)$_2M$(CF$_3$)$_4$(1,1,2-trihaloethane). Although these solvent molecules are nominally the same size, the T_cs are systematically higher for the latter because the location of the bromine atom forces the layers further apart for the 2-bromo-1,1-dichloroethane salts [29].

The effect of deuterium isotope substitution on T_c has been studied for three members of the κ_L-(BEDT-TTF)$_2$Ag(CF$_3$)$_4$(1,1,2-trihaloethane) family [12, 34–36]. In all cases, the T_c increased upon deuteration with the effect ranging from 0.21 to 0.36 K. These results are similar to the deuterium isotope effect observed in the β''-(BEDT-TTF)$_2$SF$_5$CH$_2$CF$_2$SO$_3$ [34, 37] and κ-(BEDT-TTF)$_2$Cu(NCS)$_2$ [38] superconductors.

The band electronic structure of κ_L-(BEDT-TTF)$_2$Cu(CF$_3$)$_4$(TCE) was calculated through the use of Hückel tight binding computations [39] and the infrared properties analyzed [40]. These calculations indicate that the electronic band structure [10, 41] and infrared response [42] is similar to that found in the κ-(BEDT-TTF)$_2$Cu(dca)X (X = Cl and Br) salts. Specific heat measurements of κ_L-(BEDT-TTF)$_2$Ag(CF$_3$)$_4$(TCE) indicate a linear coefficient ($\gamma = 50$ mJ mol^{-1} K^2), which is a factor of nine greater than expected from a free-electron picture [43].

2.2 κ_L'-(BEDT-TTF)$_2$Cu(CF$_3$)$_4$(DBCE)

Polymorphism has been identified in the κ_L-(BEDT-TTF)$_2$Cu(CF$_3$)$_4$(DBCE) system [44]. The primary difference between the κ_L- and κ_L'-phases is that adjacent BEDT-TTF layers in the κ_L-phase are tilted in opposite directions, while those in the κ_L'-phase are tilted in the same direction. Also similar to the κ_L-phases, the trifluoromethyl groups in the κ_L'-phase are disordered. These two phases have similar morphologies and EPR line widths [45] and are thus difficult to distinguish from each other. It is currently not known how prevalent the κ_L'-phase is, although it appears to have similar physical properties to the κ_L-phase.

2.3 κ_H-(BEDT-TTF)$_2M$(CF$_3$)$_4$(1,1,2-trihaloethane)

A needle-like morphology sometimes grows as a minority phase during the electrocrystallization process. This phase is particularly intriguing because it has notably higher T_cs (7.2–11.1 K) that approach the record for cation radical salts. (The subscript "H" designates the phase with the higher T_c.) To date, this phase has only been identified for six compositions, four of which contain the

$[Ag(CF_3)_4]^-$ anion [31]. The T_cs for the κ_H-(BEDT-TTF)$_2$M(CF$_3$)$_4$(TCE) series are 9.2 K (M = Cu) [46], 9.4 K and 11.1 K (Ag) [27] and 10.5 K (Au) [28]. Unfortunately, the filamentary nature of these crystals has, to date, prevented structural determination. The κ-phase designation has been assigned as a result of preliminary X-ray diffraction data that persistently identifies axial lengths of about 5.8 and 13 Å. These experiments hint a doubling of the unit cell along the *b*-axis [30]. EDAX indicates that the crystal composition of the κ_L- and κ_H-phases is identical with a 2:1:1 donor:anion:solvent ratio. The gradual decomposition of these crystals upon drying further indicates the presence of solvent in the κ_H-phases. Additional evidence for the presence of cocrystallized solvent comes from the variation of T_c among the κ_H-(BEDT-TTF)$_2$Ag(CF$_3$)$_4$(1,1,2-trihaloethane) series. The κ_H-phase can be readily identified by its 5–7 G EPR line width. This sharp line width is perplexing because it is much narrower than the 50–70 G associated with κ_H-phase materials. Because of the low quantity of κ_H crystals available, their fragile nature and their decomposition via solvent loss, few physical properties have been reported for this phase. However, the pressure derivative of T_c has been determined to be between -2 and -3 K kbar^{-1} [27, 28, 46], which is typical of organic superconductors [47].

2.4 *(BEDT-TTF)M(CF$_3$)$_4$ (M = Cu, Ag, Au)*

In addition to the superconducting κ-(BEDT-TTF)$_2$M(CF$_3$)$_4$(1,1,2-trihaloethane) phases, the nonsolvated (BEDT-TTF)M(CF$_3$)$_4$ phase often grows during the electrocrystallization process as rod-like crystals. The crystal structure of the silver salt has been reported [48] but severe twinning has limited structural characterization of the copper and gold salts to unit cell determinations [29]. The BEDT-TTF molecules in (BEDT-TTF)M(CF$_3$)$_4$ pack in a θ-type motif [7], which is very similar to the α-motif. In contrast to the superconducting salts, the anion's trifluoromethyl groups are ordered even at room temperature. As is typical of θ-type salts, this material is a semiconductor with an activation energy of about 0.19 eV. The room temperature EPR line width of this phase is 35–40 G, which decreases to a minimum of about 17 G at 70 K [49]. EPR has identified an uncharacterized phase transition at this temperature. EPR measurements have also demonstrated that the nonsolvated phase can be formed by heating the κ_L- and κ_H-phases to above 340 K.

2.5 *BEDT-TTF Salts Containing Derivatives of the [M(CF$_3$)$_4$]$^-$ Anion*

Anions of the type $[Ag(CF_3)_nX_{4-n}]^-$ (X = CN, CH$_3$, CCC$_6$H$_{11}$, Cl, Br and I) have been prepared [50]. The κ_L-(BEDT-TTF)$_2$Ag(CF$_3$)$_3$Cl(TCE) cation radical salt has been crystallized and was found to be isomorphous to the κ_L-(BEDT-TTF)$_2$Ag

$(CF_3)_4(TCE)$ salt [29]. However, the disorder imposed by the $[Ag(CF_3)_3Cl]^-$ anion suppresses T_c from 2.4 K to below 1.4 K. Use of the *trans*-$[Ag(CF_3)_2(CN)_2]^-$ resulted in crystallization of only the $(BEDT\text{-}TTF)_2Ag(CF_3)_2(CN)_2$ salt [29], which is isomorphous to the nonsolvated $(BEDT\text{-}TTF)_2Ag(CF_3)_4$ salt. The cyano ligands extend about 1.5 Å further from in the coordination plane from the metal center than the trifluoromethyl groups. Apparently, either geometric or electronic difference between the *trans*-$[Ag(CF_3)_2(CN)_2]^-$ and $[Ag(CF_3)_4]^-$ prevents formation of the superconducting κ_L- and κ_H-phases. The $(PPN)[Ag(CF_2H)_4]$ salt has been crystallized, structurally characterized [51] and used as an electrolyte for the attempted crystallization of cation radical salts. Surprisingly, this anion failed to produce quality crystals with BEDT-TTF, perhaps further indicating the importance of F···H interaction during the crystallization process. A BEDT-TTF salt containing the $[Cu(CF_2CF_3)_4]^-$ anion was prepared by electrocrystallization, but not structurally characterized [29]. The lack of superconductivity and the 38–46 G EPR line width suggested that this salt was related to the nonsolvated $(BEDT\text{-}TTF)_2Ag(CF_3)_4$ phase.

2.6 κ_L-$(BETS)_2Ag(CF_3)_4(TCE)$

When the BETS donor replaces the BEDT-TTF electron donor molecule during the electrocrystallization process, crystals of κ_L-$(BETS)_2Ag(CF_3)_4(TCE)$ have been prepared [29] and structurally characterized. Replacement of the inner sulfur atoms of BEDT-TTF with selenium results in a slight expansion of the unit cell and prevents the stabilization of a superconducting state above 1.2 K. Disorder in one of the BETS ethylene endgroups has been offered as a possible explanation.

3 Halophenylene Anions

3.1 Organometallic Gold(I) Anions

Crystals of $(TTFPh_2)_2[Au(C_6F_5)_2]$ and $(TTF)_2[Au(C_6F_5)_2]$ have been prepared by electrocrystallization [52, 53]. Their room temperature conductivities are 1.5 and 3 S cm^{-1}, respectively. The crystal structure of the former is characterized by segregated stacks of $TTFPh_2$ radical cations and planar $[Au(C_6F_5)_2]^-$ anions. The $[Au(C_6F_5)_2]^-$ anion has a similar geometry as present in the free anion [54]. As illustrated in Fig. 4, coordination of sulfur atoms from four $TTFPh_2$ radical cations completes an approximate octahedron about the gold center. These short Au-S contacts (3.253 and 3.490 Å) provide a structure directing entity that stabilizes the structure. An analysis of the bond lengths in the two crystallographically independent $TTFPh_2$ molecules indicates that they possess different charges.

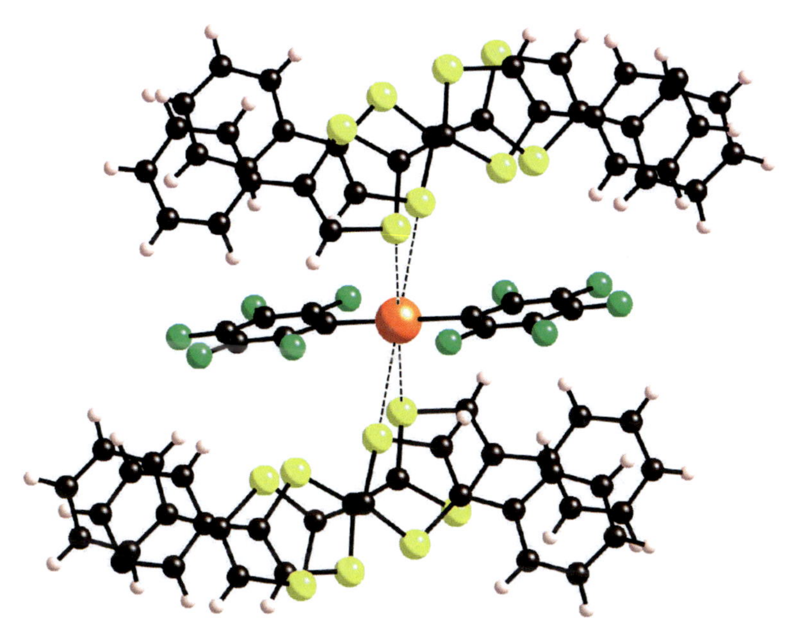

Fig. 4 Coordination environment of the gold atom in the $(TTFPh_2)_2[Au(C_6F_5)_2]$ salt

This may explain the observed semiconductive behavior with activation energy of 0.13 eV. The room temperature EPR line width of 30 G decreases to 20 G at 77 K.

Crystals of $(TTF)[Au(C_6F_5)Cl]$ have been grown by electrocrystallization [53]; however, their crystal structure has not been determined. The room temperature conductivity, as measured on compacted pellets, is quite low (10^{-6} S cm^{-1}). At room temperature, the EPR line width of these salts is about 10 G. This line width decreases with temperature as a result of increased spin-lattice relaxation times and a lower electrical conductivity.

$(TTFPh_2)[Au(C_6F_3H_2)_2]$ has been prepared as black microcrystals through the electrocrystallization method [53]. The crystal structure has not been reported, but the stoichiometry was determined through elemental analysis. This salt is characterized by a narrow 5 G EPR line width and a powder conductivity of 1.5×10^{-6} S cm^{-1}. Anions of the type $[Au(C_6X_5)_4]^-$, $[Au(C_6X_5)_2]^-$, and $[Au(C_6X_5)X]^-$ (X = F, Cl) have been used as charge compensating components in cation radical salts.

3.2 *Organometallic Gold(III) Anions*

The electrochemical process by which these salts are commonly grown has the ability to generate and stabilize new anionic species. For example, the $[Au(C_6Cl_5)_4]^-$ anion was unknown prior to the crystallization of the (BEDT-TTF)Au

$(C_6Cl_5)_4$ salt [55]. This salt formed through a complex electrochemical process involving use of (TBA)$[Au(C_6Cl_5)_2]$ as an electrolyte. Both oxidation of the metal center and intermolecular transfer of C_6Cl_5 groups occurred to generate the $[Au(C_6Cl_5)_4]^-$ anion, which was subsequently crystallized with the $(BEDT\text{-}TTF)^+$ cation. In this salt, the $(BEDT\text{-}TTF)^+$ cations are completely surrounded by the pentachlorophenyl rings of $[Au(C_6Cl_5)_4]^-$, providing an ideal opportunity to study a highly isolated BEDT-TTF cation. Attempts to crystallize the related (BEDT-TTF)$Au(C_6F_5)_4$ salt through use of electrolytes containing either the $[Au(C_6F_5)_4]^-$ or $[Au(C_6F_5)_2]^-$ anions have been unsuccessful [55].

The $(TTFPh)_{2.5}[Au(C_6F_5)_2Cl_2]$, (TTFPh)$[Au(C_6F_5)_2I_2]$ and (TTF)$[Au(C_6F_3H_2)_2Cl_2]$ salts have been prepared by electrocrystallization [53]. Their crystal structures have not been determined but their stoichiometries have been estimated through elemental analysis. Their room temperature conductivities, as measured on compressed pellets, are 2×10^{-4}, 1×10^{-3}, and 2×10^{-6} S cm^{-1}, and their EPR line width are 7, 5, and 10 G, respectively.

3.3 *Perfluoro-ortho*-Phenylene Mercury Complex

The cocrystal adduct $TTF[Hg_3(C_6F_4)_3]_2$ crystallizes as orange needles by combining 1:1 carbon disulfide: methylene chloride solutions of TTF and $Hg_3(C_6F_4)_3$ [56]. As illustrated in Fig. 5, the crystal structure is stabilized by multiple Hg···S secondary interactions which cause the TTF molecules to be sandwiched between two $Hg_3(C_6F_4)_3$ molecules. Spectroscopic and structural results indicate that charge transfer does not occur in this adduct and minimal conductivity is expected.

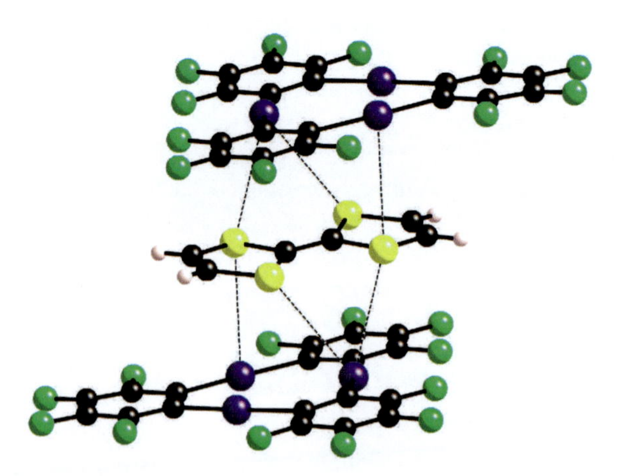

Fig. 5 Sandwich adduct of $TTF[Hg_3(C_6F_4)_3]_2$ illustrating Hg···S secondary interactions which are less than 3.6 Å

4 Metallacarborane Salts

A series of TTF complexes containing metallacarboranes have been reported. One motivation for this work was to prepare materials that are complementary to the decamethylmetallocenium salts that exhibit cooperative magnetic phenomena [57, 58]. These materials also have potential for the formation of hybrid charge transfer salts that combine magnetic and conductive properties. Salts of the type $(TTF)^+[M(C_2B_9H_{11})_2]^-$ (M = Cr, Fe and Ni) were prepared by metathesis reactions [59]. The geometry of the TTF cation and the infrared spectra of these salts are consistent with an oxidation state of 1+. For the chromium salt, the crystal structure is characterized as alternating layers of $(TTF)^+$ cations and $[Cr(C_2B_9H_{11})_2]^-$ anions with no close intermolecular S···S contacts in the TTF layer [60]. The structures of the iron and nickel analogs are characterized by discrete stacks of $(TTF)^+$ dimers. These materials behave as simple paramagnets and cooperative magnetic behavior is not observed. However, these salts show considerable variations in their conduction properties as the number of *d* electrons in the carborane is changed: the chromacarborane is a semiconductor with a room temperature conductivity of 3×10^{-4} S cm^{-1} and an activation energy of 0.16 eV, while the nickelacarborane derivative is an insulator [61].

These significant differences in the conduction properties of the $(TTF)^+[M(C_2B_9H_{11})_2]^-$ (M = Cr and Ni) salts prompted the investigation of related carborane salts in which a pendant thiophene group had been added in order to promote mixed valency in the TTF stacks [61, 62]. Two TTF salts have been reported using this approach: $(TTF)[Fe(C_2B_9H_{10}C_4H_3S)_2]$(toluene) (1:1) and $(TTF)_5[Fe(C_2B_9H_{10}C_4H_3S)_2]_2$ (5:2) [61, 62]. The crystal structure of the 1:1 salt is characterized by discrete stacks of $(TTF)^+$ dimers which results in insulating electrical properties. In contrast, the crystal structure of the 5:2 salt contains an unusual two-dimensional TTF network in which stacks of TTF trimers are linked to orthogonal TTF monomers through short S···S contacts. Variations in the bond lengths of the crystallographically inequivalent TTF molecules and an intense, broad charge-transfer band in the infrared spectrum suggest mixed valency. The semiconductive (activation energy 0.22 eV) 5:2 salt shows significantly higher room temperature conductivity (2×10^{-3} S cm^{-1}) than the 1:1 salt. Magnetically, both salts exhibit Curie-Weiss behavior with weak ferromagnetic interactions.

The BEDT-TTF salts of two paramagnetic metallacarborane anions have been reported: the $(BEDT\text{-}TTF)_2Cr(C_2B_9H_{11})_2$ salt was prepared by chemical oxidation methods while the $(BEDT\text{-}TTF)_2Fe(C_2B_9H_{10}C_4H_3S)_2$ salt was prepared by electrocrystallization [63]. The packing motif of the BEDT-TTF molecules in $(BEDT\text{-}TTF)_2Cr(C_2B_9H_{11})_2$ is of the β' type (slipped stacks of BEDT-TTF dimers) [6] and the formal charge of the BEDT-TTF molecules, based on a bond-length analysis [64], was determined to be +0.5. The crystal structure of $(BEDT\text{-}TTF)_2Fe(C_2B_9H_{10}C_4H_3S)_2$ is characterized by a loosely packed honeycomb network of short intermolecular S···S interactions. It is interesting to note that S···S interactions between the BEDT-TTF molecules are favored over S···S interactions between

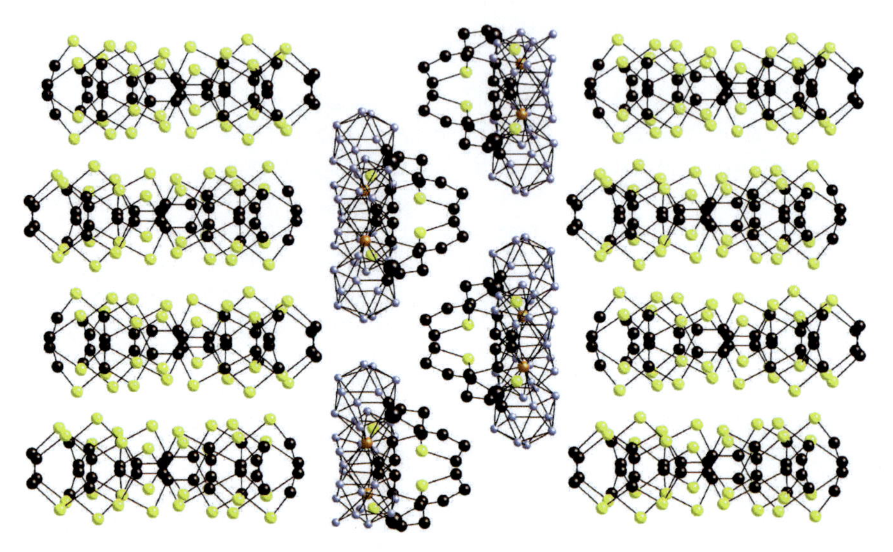

Fig. 6 A view down the *c*-axis of the (BEDT-TTF)Fe($C_2B_9H_{10}C_4H_3S$)$_2$ salt, illustrating the very thick anionic layers which result from the thiophene derivatized iron carborane anions. Hydrogen atoms and disordered thiophene groups have been omitted for clarity

BEDT-TTF and the thiophene moiety of the anion. As illustrated in Fig. 6, the crystal structure of this salt contains double anion layers resulting in a very large separation (~12 Å) of the BEDT-TTF layers. Both BEDT-TTF salts exhibit semiconductive electrical properties with a room temperature compressed pellet conductivity of 2×10^{-3} S cm^{-1} for (BEDT-TTF)$_2$Cr($C_2B_9H_{11}$)$_2$ and a single crystal conductivity of 5×10^{-1} S cm^{-1} for (BEDT-TTF)$_2$Fe($C_2B_9H_{10}C_4H_3S$)$_2$. Both salts exhibit Curie-Weiss behavior with weak antiferromagnetic and ferromagnetic interactions for (BEDT-TTF)$_2$Cr($C_2B_9H_{11}$)$_2$ and (BEDT-TTF)$_2$Fe ($C_2B_9H_{10}C_4H_3S$)$_2$, respectively.

5 Metallocenium Complexes

5.1 Metallocenium Anions

Metallocenium charge-transfer complexes have been widely studied for their novel magnetic properties [65]. The first metallocenium salts of TTF derivatives were prepared by exploiting the strongly oxidizing nature of (MeCp)VCl_3 [66]. When a 1:1 molar ratio of (MeCp)VCl_3 and TMTTF were combined in a methylene chloride solution, the (TMTTF)[(MeCp)VCl_3] salt was formed. In this crystal structure, TMTTF cations are strongly dimerized and assemble to form a zigzag chain. The $[(MeCp)VCl_3]^-$ anions are located in cationic pockets defined by the TMTTF dimers. By tuning the crystallization conditions, it was possible to crystallize the $(TMTTF)_3[[(MeCp)VCl_2]_2(\mu\text{-}O)]_2$ salt. This salt contains a μ-O bridged anion similar to that found in $Cp_2^*V_2I_4O$ [67]. The crystal structure is characterized by partially oxidized stacks of TMTTF molecules separated by sheets of mixed-valence bimetallic anions. Resistivity measurements indicated that this salt is a semiconductor with a room temperature conductivity of 5.1×10^{-6} S cm^{-1} and an activation energy of 0.54 eV. A ferromagnetic transition was reported to occur at 20 K.

Crystals of α-$(BEDT\text{-}TTF)_4[Fe(Cp\text{-}CONHCH_2SO_3)_2]\cdot 4H_2O$ have been prepared by electrocrystallization through the use of tetraphenylphosphonium salt of the derivatized ferrocenium anion [68]. The crystal structure of this cation-radical salt is characterized by cationic layers that contain two types of BEDT-TTF tetramers. Overall, the packing motif is of the α-type [7]. The molecular geometry of the BEDT-TTF molecules suggest that charge disproportionation occurs within the structure. This salt displays semiconductive properties with a relatively high room temperature conductivity of 0.16 S cm^{-1} and an activation energy of 0.11 eV. Because the ferrocenyl moiety is in the neutral state the salt is not significantly magnetic. More recently, the β''-$(BEDT\text{-}TTF)_4[Fe(Cp\text{-}CONHCH_2SO_3)_2]\cdot 2H_2O$ polymorph of this compound has been crystallized by an analogous electrocrystallization procedure [69]. An analysis of the bond-lengths of the two crystallographically independent BEDT-TTF molecules indicates that both have a similar +0.5 charge. These radical cations pack in a two-dimensional network consisting of stacks of canted BEDT-TTF molecules, typical of a β″-packing motif [6]. Electrical conductivity measurements on single crystals demonstrate that the material is metallic from room temperature down to 70 K, at which point a metal-to-insulator (MI) transition occurs. These results are consistent with electronic band structure calculations. The ferrocenyl moiety is again in the neutral state and thus the small and temperature-independent magnetic susceptibility is attributed to the BEDT-TTF radical cations. Application of hydrostatic pressure up to 9.5 kbar suppresses the MI transition to 1.6 K; however, the MI transition reappears at pressures above 11.5 kbar as a result of a proposed pressure induced phase transition.

The 1,1′-ferrocenedisulfonate anion has been used as a charge compensating anion in the α'''-$(BEDT\text{-}TTF)_4[Fe(C_5H_4SO_3)_2]\cdot 6H_2O$ salt [70]. As illustrated in Fig. 7, the novel electron-donor layer motif is a combination of the α- and β″-packing motifs. An analysis of the molecular structure indicates that the BEDT-TTF molecules have an average formal charge of +0.5, but the charges on the crystallographically independent molecules range from +0.37 to +0.79. The ferrocenyl part is in the neutral state. This material behaves as a semiconductor with a room temperature conductivity of $\sim 8 \times 10^{-2}$ S cm^{-1} and an activation energy of 0.22 eV.

TTF salts containing similar ferrocenium anions have been prepared through metathesis reactions between $(TTF)_3(BF_4)_2$ and the tetraphenylphosphonium salts of the desired anions [71]. The crystal structure of $(TTF)(CpFeCp\text{-}CONHCH_2SO_3)\cdot 1/3H_2O$ contains isolated TTF dimers. Hydrogen bonds of the –NH···OS– and –NH···O=C– type are present between anions. Two phases of the TTF salts with the

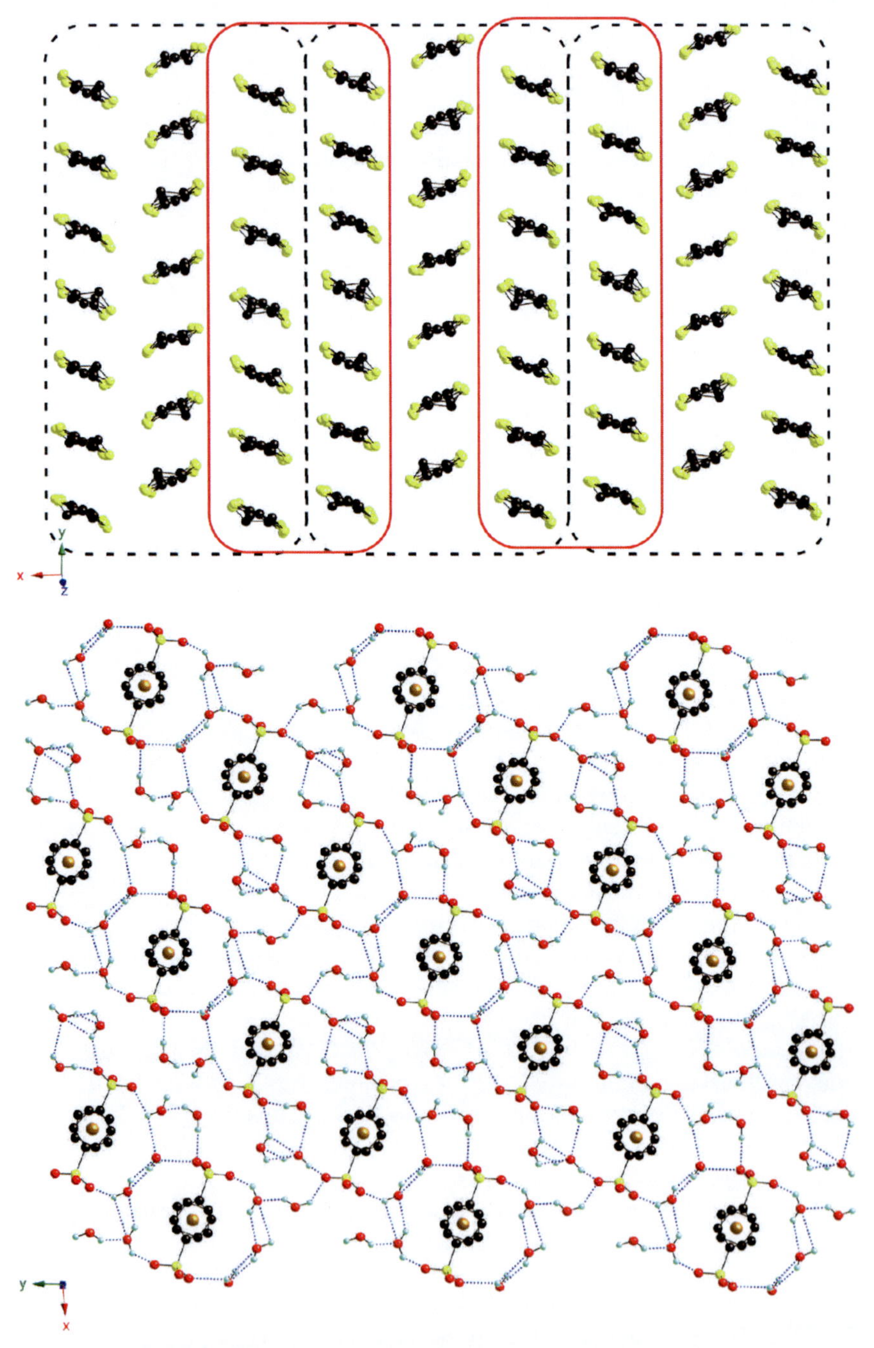

Fig. 7 The novel packing motif of α‴-(BEDT-TTF)4([Fe($C_5H_4SO_3$)$_2$]·6H_2O. Regions containing α-type circled with a *dotted line*, while regions with β″-packing are highlighted with an *oval* (*top*). The anionic layer is characterized by an extensive hydrogen bonding network (*dashed bonds*) between water molecules and the sulfonate functionality of the anion

$[Fe(Cp\text{-}CONHCH_2SO_3)_2]^-$ anion have been identified: $(TTF)_2[Fe(Cp\text{-}CONHCH_2SO_3)_2]$ and the solvated $(TTF)_2[Fe(Cp\text{-}CONHCH_2SO_3)_2]\cdot 2CH_3OH$. Both crystal structures contain face-to-face dimers, with the main structural differences probably attributed to hydrogen bonding to the cocrystallized methanol. No physical properties of these salts have yet been published.

5.2 Covalently Attached Ferrocene Derivatives of Tetrathiafulvalene

As illustrated in Scheme 4, TTF derivatives have been prepared in which ferrocenium moieties have been covalently attached. The synthetic aspects of this work have recently been reviewed [72]. One motivation for this work is to control electronic spin and transport through the coupling of localized (magnetic) and iterant (conduction) electrons in a single system. The first research in this direction was reported more than a quarter of a century ago with the preparation of diferrocenyltetrathiafulvalene [73]. This molecule contains two electron donor moieties: TTF and ferrocene. The 1:1 charge transfer complexes with TCNQ and DDQ have been obtained as blackish green solids. The polycrystalline room temperature electrical conductivity of the DDQ salt is $\sim 1.2 \times 10^{-3}$ S cm^{-1}. These salts have not been structurally characterized. TTF-Li can be reacted with acetylferrocene to produce TTF-CMe(OH)Fc, which through subsequent treatment with methyl iodide produces TTF-CMe(OMe)Fc [74]. Hydrogen bonding is important in controlling the solid state structure of TTF-CMe(OH)Fc, resulting in a packing motif in which the ferrocene moiety of each molecule is sandwiched between TTF moieties. However, in the case of TTF-CMe(OMe)Fc, hydrogen bonding is less important and a crystal structure containing dimerized TTF moieties results.

The 1-tetrathiafulvalenylferrocene electron donor molecule has also been synthesized [75]. The charge transfer salts formed by chemical oxidation with TCNQ, DDQ and $TCNQF_4$ proved to be insulating. Cation radical salts with ClO_4^-, PF_6^-, BF_4^-, AuI_2^-, I_3^-, IBr_2^-, and I_2Br^- were prepared by electrocrystallization and also found to be insulating. The salts with $M(dmit)_2^-$ (M = Ni, Pd and Pt) were found to be semiconductive with a room temperature conductivity of about 10^{-6} S cm^{-1}. As illustrated in Fig. 8, the (1-tetrathiafulvalenylferrocene) $Ni(dmit)_2$ salt has been structurally characterized and shown to consist of cofacial $TTF/Ni(dmit)_2$ units.

The 1,1′-diferrocenyl-VT electron donor molecule is structurally similar to diferrocenyltetrathiafulvalene but with the TTF moiety replaced by bis(vinylenedithio)tetrathiafulvalene (VT) [76]. It has currently not been possible to separate the *cis*- and *trans*-isomers. The 1:1 polyiodide complex of 1,1′-diferrocenyl-VT was obtained through reaction with iodine. EPR and Mössbauer spectra indicate that in this charge transfer salt the VT moiety is oxidized while the ferrocene

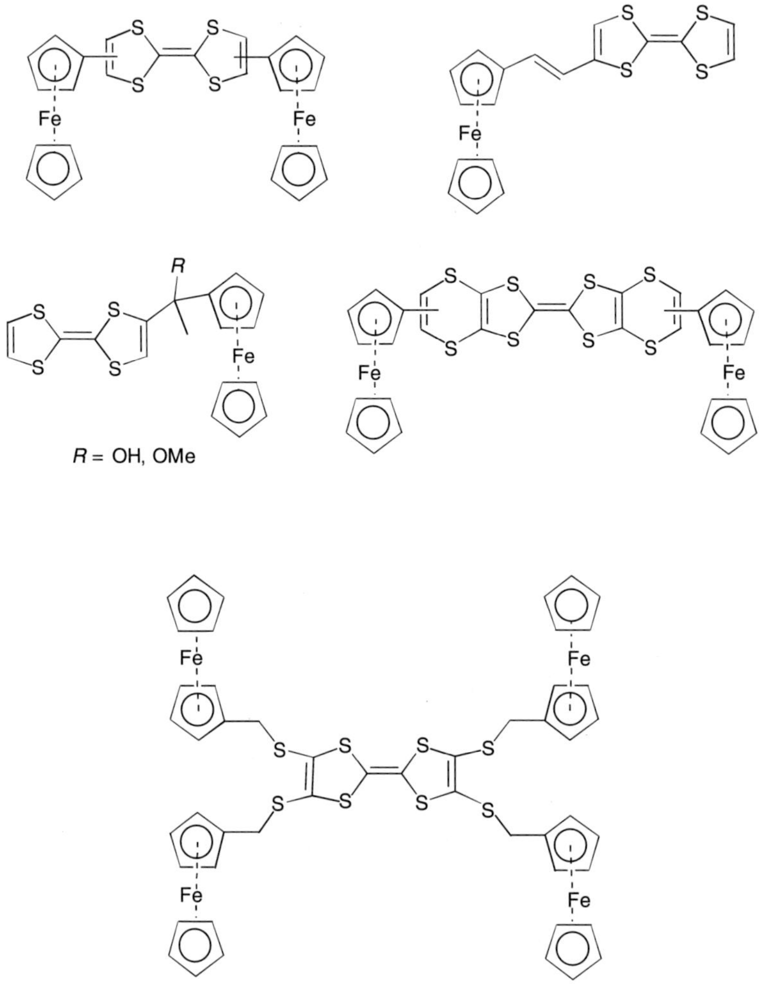

Scheme 4 Derivatives of TTF in which the ferrocenium functionality has been covalently attached

units remain neutral. The compressed pellet conductivity of this complex is about 7×10^{-3} S cm^{-1} at room temperature.

Over the past decade a number of new covalently bonded TTF/ferrocene adducts have been reported [77, 78]. The crystal structure of the 1,1′-bis(1,3-dithiole-2-ylidine)-substituted ferrocene derivative has been published [77]. In this complex, ferrocene has essentially been incorporated as a molecular spacer between the two 1,3-dithole-2-ylidene rings forming a stretched TTF molecule. This adduct, and its methyl-substituted derivative, have been combined with TCNQ to form charge-transfer complexes with room temperature powder conductivities of ~0.2 S cm^{-1}. Similar diferrocenyl complexes have been prepared with bis (dithiolene) metal complexes [79, 80].

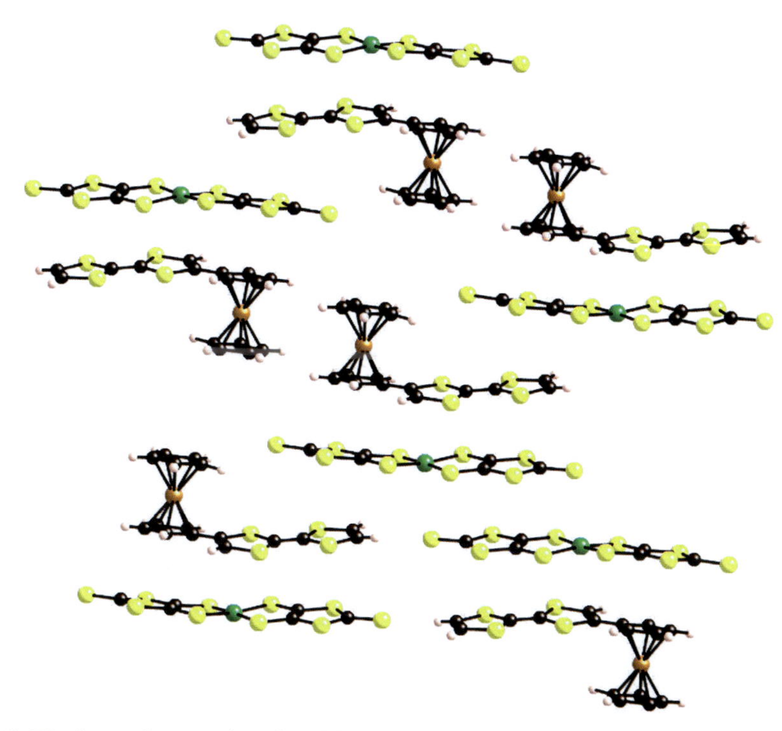

Fig. 8 The face-to-face stacking of the TTFFc and $Ni(dmit)_2$ molecules in the $(TTFFc)Ni(dmit)_2$ crystal structure

5.3 *Dithiolate Complexes Containing Cyclopentyldienyl Ligands*

The $Rh(\eta^5\text{-}C_5H_5)(C_8H_4S_8)$ and $M(\eta^5\text{-}C_5Me_5)(C_8H_4S_8)$ (M = Rh and Ir) complexes (see Scheme 5.) have been prepared by the reactions of $(TMA)_2(C_8H_4S_8)$ with $[Rh(\eta^5\text{-}C_5H_5)Cl_2]_2$ or $[M(\eta^5\text{-}C_5Me_5)Cl_2]_2$ (M = Rh and Ir) [81, 82]. The one-electron-oxidized species $[M(L)(C_8H_4S_8)]^+$ and $[MX(L)(C_8H_4S_8)]$ (M = Rh and Ir; X = Br and I; L = $\eta^5\text{-}C_5H_5$ and $\eta^5\text{-}C_5Me_5$) have been prepared via oxidation with bromine, iodine or the ferrocenium cation. The EPR spectra indicate that the oxidation occurs essentially on the sulfur-rich dithiolate ligand. The two-electron-oxidized species $[MI(\eta^5\text{-}C_5H_5)(C_8H_4S_8)](I_3)$ (M = Rh and Ir) have also been prepared and their crystal structures determined. The solid state structure contains many S···S and I···S contacts that link the cations into a 2D array.

The $[MX(\eta^5\text{-}C_5Me_5)(C_8H_4S_8)]$ and $[MI(\eta^5\text{-}C_5Me_5)(C_8H_4S_8)]A$ (M = Rh and Ir; X = Br and I; A = I_3 and PF_6) complexes have room temperature conductivities between 10^{-7} and 2×10^{-4} S cm^{-1} [81]. The poor conductivity of these materials has been attributed to the steric effects of the $\eta^5\text{-}C_5Me_5$ ligand that minimizes the effectiveness of the S···S contacts as conduction pathways. The $[IrI(\eta^5\text{-}C_5Me_5)$

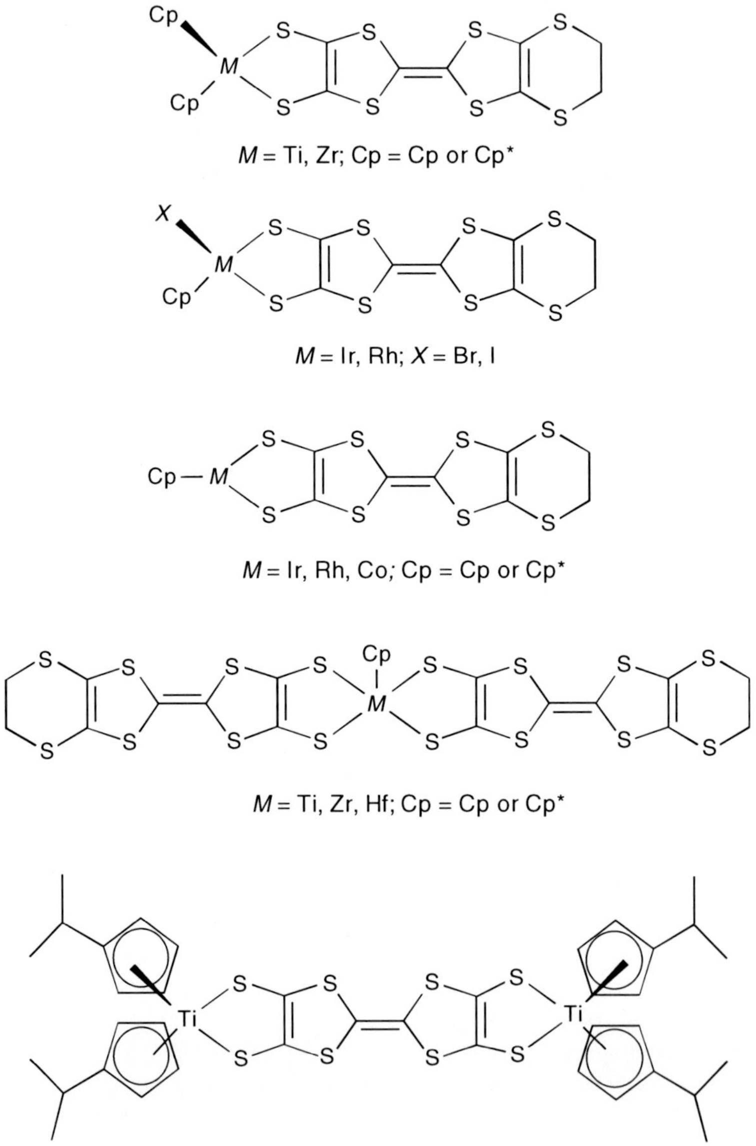

M = Ti, Zr; Cp = Cp or Cp*

M = Ir, Rh; X = Br, I

M = Ir, Rh, Co; Cp = Cp or Cp*

M = Ti, Zr, Hf; Cp = Cp or Cp*

Scheme 5 Dithiolate complexes containing cyclopentyldienyl ligands

$(C_8H_4S_8)](I_3)_{1/2}(I_7)_{1/2}$ salt has a compressed pellet room temperature conductivity of 1.2×10^{-6} S cm^{-1} [82]. The presence of the $(I_7)^-$ anion is confirmed by a single crystal structure analysis. Nonbonded S···S and S···I contacts result in the formation of a layer structure. The $(I_7)^-$ anions are components of this layer, while the $(I_3)^-$ anions reside between layers. In contrast, the $[Rh(\eta^5\text{-}C_5H_5)(C_8H_4S_8)]X$ (X = Br and I) have notably higher conductivities of 2.5×10^{-2} S cm^{-1} (X = I) and 6.5×10^{-2} S cm^{-1} (X = Br).

The related complex, $Co(\eta^5\text{-}C_5H_5)(C_8H_4S_8)$ has been prepared [83]. The $[Co(\eta^5\text{-}C_5H_5)(C_8H_4S_8)](I_3)$ and $[Co(\eta^5\text{-}C_5H_5)(C_8H_4S_8)](PF_6)_{0.7}$ salts, which were formed by chemical oxidation, have room temperature compressed pellet conductivities of 0.19 and 0.16 S cm^{-1}, respectively.

The following cyclopentadienyl titanium(IV) complexes containing the $(C_8H_4S_8)^{2-}$ ligand have been synthesized: $Ti(\eta^5\text{-}C_5H_5)_2(C_8H_4S_8)$, $Ti(\eta^5\text{-}C_5Me_5)_2(C_8H_4S_8)$, $(TMA)[Ti(\eta^5\text{-}C_5H_5)(C_8H_4S_8)_2]$ and $(TMA)[Ti(\eta^5\text{-}C_5Me_5)(C_8H_4S_8)_2]$ [84]. As synthesized, these complexes are electrical insulators, but a variety of cation radical salts have been prepared through chemical oxidation with iodine, ferrocenium, or TCNQ. The conductivity of these oxidized complexes has been measured on polycrystalline pellets at room temperature. The complexes with two bulky cyclopentadienyl rings have conductivities of $\sim 10^{-4}$ S cm^{-1}. In contrast, those with two dithiolate moieties presumably exhibit enhanced S···S interactions which result in conductivities of about 0.1 S cm^{-1}. The crystal structure of $Ti(\eta^5\text{-}C_5Me_5)_2(C_8H_4S_8)$ shows pseudotetrahedral coordination of the titanium(IV) ion. The molecules pack in a dimerized structure with intradimer S···S contacts.

A related series of materials have been prepared in which titanium is replaced with zirconium and hafnium: $Zr(\eta^5\text{-}C_5H_5)_2(C_8H_4S_8)$, $(TMA)[Zr(\eta^5\text{-}C_5H_5)(C_8H_4S_8)_2]$, $(TMA)[Zr(\eta^5\text{-}C_5Me_5)(C_8H_4S_8)_2]$ and $(TMA)[Hf(\eta^5\text{-}C_5Me_5)(C_8H_4S_8)_2]$ [85]. As in the case of titanium, oxidation results in a dramatic increase in conductivity, with polycrystalline room temperature conductivities as high as 0.1 S cm^{-1} for $[Zr(\eta^5\text{-}C_5H_5)_2(C_8H_4S_8)](I_3)$. Raman spectroscopy has identified the form of the iodide in this complex to be triiodide. The crystal structures of the zirconium and hafnium complexes have not been determined, thus precluding a thorough analysis of the conduction mechanism.

The first example of a homobimetallic tetrathiafulvalene tetrathionate complex was described a little over a decade ago [86, 87]. The crystal structure of $(i\text{-}PrC_5H_4)_2Ti[S_2TTFS_2]Ti(i\text{-}PrC_5H_4)_2 1.5C_6H_6$ shows that the molecular structure consists of a planar $TTFS_4$ core with one of the peripheral $Ti(i\text{-}PrC_5H_4)_2$ groups situated above this plane and the other below. The central TTF core has essentially the same bond lengths and angles as neutral TTF, thus indicating that there is no internal oxidation of the TTF by titanium. These molecules pack in a unique layer motif, in which the bimetallic TTF molecules are arranged perpendicular to each other. This motif is related to the τ-motif in BEDT-TTF salts.

6 Metal Carbonyls

6.1 Diphenylphospino Complexes

As illustrated in Scheme 6, the TTF core has been functionalized with diphenylphosphino moieties to form derivatives such as 3,4-dimethyl-3′,4′-bis (diphenylphosphino)tetrathiafulvalene [$o\text{-}Me_2TTF(PPh_2)_2$] [88] and tetrakis

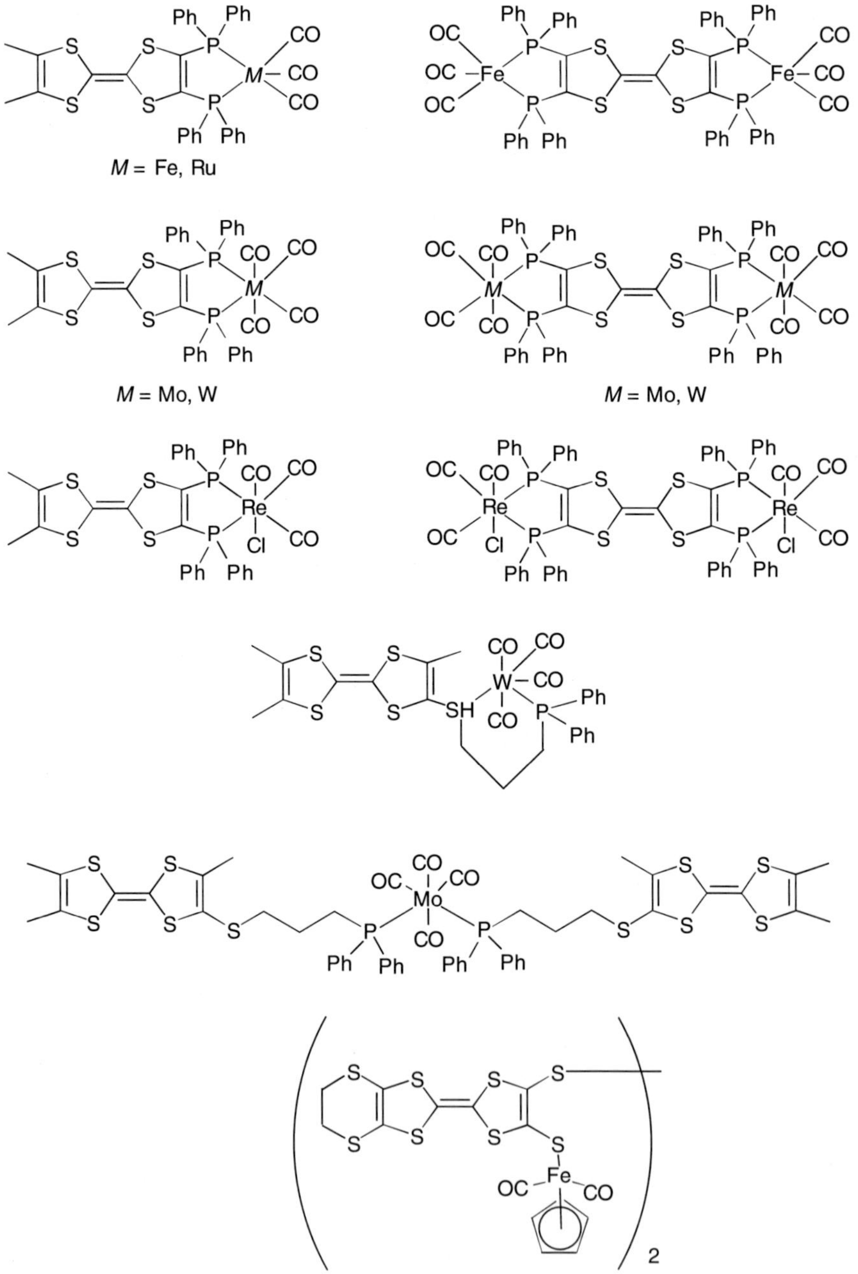

M = Fe, Ru

M = Mo, W

M = Mo, W

Scheme 6 Carbonyl complexes containing diphenylphosphino derivatives of TTF

(diphenylphosphino)teterthiafulvalene [$TTF(PPh_2)_4$] [89]. Organometallic Mo $(CO)_4$, $W(CO)_4$, $M(CO)_3$ (M = Fe and Ru), and $Re(CO)_3Cl$ complexes have been formed with these electron-donor molecules [90, 91]. The coordination geometry of the various products has been studied through the use of infrared, ^{31}P-NMR and ^{13}C-NMR spectroscopies. The crystal structures of the $[TTF(PPh_2)_4][W(CO)_4]_2$, [*o*-$Me_2TTF(PPh_2)_2$]$Re(CO)_3Cl$, [*o*-$Me_2TTF(PPh_2)_2$]$M(CO)_3$ (M = Fe and Ru), and $[TTF(PPh_2)_4][Fe(CO)_3]_2$ complexes have been determined by single crystal diffraction. The tungsten, molybdenum, and rhenium complexes exhibit octahedral coordination. In the case of $[TTF(PPh_2)_4]Re(CO)_3Cl$, the carbonyl ligands are arranged in *fac* orientation. Two peaks in the ^{31}P-NMR spectra indicate the presence of two diastereoisomers resulting from the chloride ligand adopting a position either *cis* or *trans* to the TTF plane. The iron and ruthenium complexes display easy interconversion between trigonal bipyramidal and square planar pyramidal geometry in solution, but reside in a distorted trigonal bipyramidal coordination in the solid state.

The redox properties of the [*o*-$Me_2TTF(PPh_2)_2$][$M(CO)_3$] (M = Fe and Ru) have been studied in great detail [91]. One-electron oxidation of [*o*-$Me_2TTF(PPh_2)_2$][Ru $(CO)_3$] results in decomposition of the complex with formation of [*o*-Me_2TTF $(PPh_2)_2]^+$. In contrast, one-electron oxidation of [*o*-$Me_2TTF(PPh_2)_2$][$Fe(CO)_3$] results in the formation of [*o*-$Me_2TTF(PPh_2)_2$][$Fe^{(I)}(CO)_3$] through metal-centered oxidation. This oxidized species is only modestly stable and decomposes by scission of a P-Fe bond. If free ligand is present during this process, the [*o*-$Me_2TTF(PPh_2)_2$]$_2$[$Fe^{(I)}(CO)$] species forms.

Electrocrystallization has been used to grow single crystals of the {[*o*-Me_2TTF $(PPh_2)_2$][$W(CO)_4$]$_2$}$_2$[Mo_6O_{19}] cation radical salt [92]. An analysis of the bond lengths in the solid state structure indicate that the oxidation occurs on the TTF moiety. In order to balance the charge of the $[Mo_6O_{19}]^{2-}$ anion, the TTF moiety is oxidized to the +1 state. As illustrated in Fig. 9, the {[*o*-$Me_2TTF(PPh_2)_2$] $[W(CO)_4]_2\}^+$ cations form well isolated uniform chains. The magnetic susceptibility has been successfully fitted as a Heisenberg spin chain with $J/k = -17$ K. Infrared spectroscopy indicates that the carbonyl absorptions occur at higher frequencies for the oxidized complex, a result of weaker π-back donation upon oxidation.

As illustrated in Scheme 6, the 3-[3-(diphenylphosphino)propylthio]-3′,4,4′-trimethyl-tetrathiafulvalene ligand can act as either a monodentate [in the case of $Mo(CO)_4$] or bidentate [in the case of $W(CO_4)$] ligand for the formation of metal carbonyl complexes [93]. Even though the mode of coordination is different in these two complexes, the solid state packing motif is characterized by alternate layers of organic (TTF) and inorganic moieties. The proximity of the TTF core weakly influences the redox properties of the metal center for the case of the bidentate tungsten complex.

The dinuclear $[Fe(C_5H_5)(CO)_2(C_8H_4S_8\text{-}C_8H_4S_8)Fe(C_5H_5)CO_2]$ organometallic complex exhibits ligand-centered oxidation at a low potential [94]. The crystalline material that forms upon chemical oxidation with iodine contains both I_3^- and I_5^- anions and has a room temperature compressed pellet conductivity of 1.7×10^{-4} S cm^{-1}.

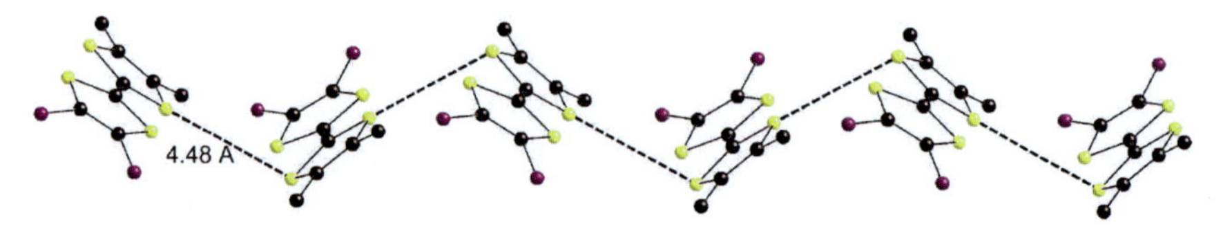

Fig. 9 Well isolated spin chain of $\{[o\text{-}Me_2TTF(PPh_2)_2][W(CO)_4]_2\}^+$ cations in the $\{[o\text{-}Me_2TTF(PPh_2)_2][W(CO)_4]_2\}[Mo_6O_{19}]$ cation radical salt. Hydrogen atoms, phenyl rings, $[Mo_6O_{19}]^{-2}$ anions and CO groups have been omitted for clarity

6.2 *Salts of Anionic Metal Carbonyls*

Single crystals of $(TTF)Rh(CO)_2Cl_2$ have been prepared by combining acetonitrile solutions of $Rh_2(CO)_4Cl_2$, (TEA)Cl and $(TTF)_3(BF_4)_2$ [95]. The crystal structure is characterized by a checker board arrangement of segregated stacks of $(TTF)^{+1}$ cations and $[Rh(CO)_2Cl_2]^-$ anions. The square planar anions structurally disordered, but based on infrared analysis [96], are believed to be in the *cis* configuration. This salt exhibits semiconductive properties with a room temperature conductivity of 2.1×10^{-4} S cm^{-1} and an activation energy of 0.26 eV.

7 *C*-Deprotonated-2-phenylpyridine(−) Derivatives

As illustrated in Scheme 7, the *C*-deprotonated-2-phenylpyridine (ppy) ligand has been used as a component of (ppy)M(S-S) (M = Au^{3+}, Pt^{2+}; $S\text{-}S^{2-} = C_8H_4S_8{}^{2-}$, $C_8H_4S_6O_2{}^{2-}$) organometallic complexes that contain a TTF moiety. This ligand forms strong M-C σ-bonds that can be used to tune the electron-donating abilities and band fillings of its salts.

The synthesis of $Au(ppy)(C_8H_4S_8)$ and $Au(ppy)(C_{10}\text{-}C_6S_8)$, including the crystal structure of the former, have been reported [97]. The triiodide and TCNQ salts of both complexes have also been prepared by chemical oxidation. Analysis of the EPR spectra indicates that the oxidation is centered on the dithiolate ligands. High room temperature conductivities of $2\text{–}4 \times 10^{-2}$ S cm^{-1} were measured on compacted polycrystalline samples for the oxidized complexes.

Electrocrystallization has been used to grow crystals of $[Au(ppy)(S\text{-}S)]_2[Q][Solvent]_n$ (S-S = $C_8H_4S_8$ and $C_8H_4S_6O_2$; $Q = PF_6{}^-$, $BF_4{}^-$, $AsF_6{}^-$ and $TaF_6{}^-$; Solvent = PhCl and PhCN; n = 0–0.5) [98]. The crystal structures of $[Au(ppy)(C_8H_4S_8)]_2[PF_6]$ and $[Au(ppy)(C_8H_4S_6O_2)]_2[BF_4]$ are characterized by stacks of partially oxidized $[Au(ppy)(C_8H_4S_8)]^{0.5+}$ cations that are linked into layers through short S···S contacts [98, 99]. As illustrated in Fig. 10, the intrastack packing is influenced by S···S contacts in the former while O···HC hydrogen bonds in the latter cause slipping of the electron donor molecules preventing formation of intrastack S···S contacts. It is interesting to note that the O···HC hydrogen bonds that

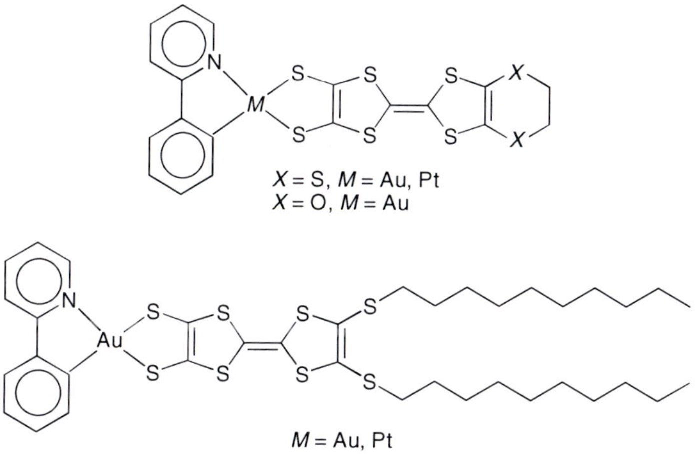

$X = S$, $M = Au$, Pt
$X = O$, $M = Au$

$M = Au$, Pt

Scheme 7 TTF-based molecules that contain the *C*-deprotonated-2-phenylpyridine (ppy) ligand

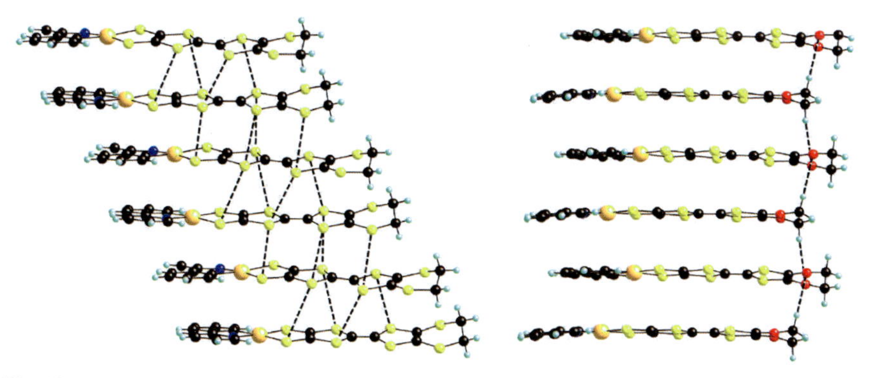

Fig. 10 Side view of the cation radical stacks in the $[Au(ppy)(C_8H_4S_8)]_2[PF_6]$ (*left*) and $[Au(ppy)(C_8H_4S_6O_2)]_2[BF_4]$ (*right*) salts. The packing motif is influenced by S···S contacts in the $[Au(ppy)(C_8H_4S_8)]_2[PF_6]$ salt and O···HC hydrogen bonds in $[Au(ppy)(C_8H_4S_6O_2)]_2[BF_4]$

apparently stabilize the insulating state in $[Au(ppy)(C_8H_4S_6O_2)]_2[BF_4]$ are well known to stabilize a metallic state in many BEDO-TTF cation radical salts [100]. The crystal structures of $[Au(ppy)(C_8H_4S_8)]_4[XF_6]_2[PhCl]$ (X = As and Ta) are characterized by alternate stacking of tetramers.

The $[Au(ppy)(C_8H_4S_8)]_4[XF_6]_2[PhCl]$ (X = As and Ta) salts exhibit semiconductive behavior, with ($\sigma_{RT} = 2 \times 10^{-3}$–$2 \times 10^{-4}$ S cm^{-1}, $E_a = 0.11$–0.15 eV). The insulating behavior ($\sigma_{RT} = 10^{-5}$ S cm^{-1}) of the $[Au(ppy)(C_8H_4S_6O_2)]_2[BF_4]$ salt is a result of dimerization and is understood through electronic band structure calculations [101, 102]. The room temperature conductivity of $[Au(ppy)(C_8H_4S_8)]_2[PF_6]$ is 9.2 S cm^{-1} in the stacking direction and 0.0028 S cm^{-1} in the interlayer direction. At ambient pressure, semiconductive behavior is observed with an activation energy of 0.03 eV. The first example of a metallic cation radical salt based on unsymmetrical mixed-ligand organometallic complex was realized through application of 0.8 GPa of pressure [98] with metallic behavior to as low as 20 K under application of 2 GPa [98]. Magnetic susceptibility measurements indicate that $[Au(ppy)(C_8H_4S_8)]_2[PF_6]$ is a Mott insulator close to the metal-insulator boundary [102].

Related platinum complexes, $(TBA)[Pt(ppy)(C_8H_4S_8)]$ and $(TEA)[Pt(ppy)(C_{10}\text{-}C_6S_8)]$, have also been reported [103]. Chemical oxidation with TCNQ afforded the neutral [Pt(ppy)L] complex, while oxidation with iodine produced [Pt(ppy)L] $(I_{3.6\text{–}4.2})$ (L = $C_8H_4S_8$ and $C_{10}\text{-}C_6S_8$). The form of iodide in these complexes has been identified as a combination of I_3^- and I_5^- by Raman spectroscopy. EPR spectroscopy indicates that the oxidation is dithiolate ligand centered. The compressed pellet conductivities of the oxidized species were 6.2×10^{-6}–7.8×10^{-3} S cm^{-1}. The single crystal structure of the $(TBA)[Pt(ppy)(C_8H_4S_8)]$ salt has been reported [103], revealing that the $[Pt(ppy)(C_8H_4S_8)]^-$ anion has planar geometry around the metal center but a 32° bending of the $C_8H_4S_8$ ligand.

8 Cocrystals

One example of a cocrystal containing BEDT-TTF and ferrocene has been reported [104]. The $[(BEDT\text{-}TTF)_2C_{60}]_2{\cdot}$(ferrocene) adduct was crystallized from a carbon disulfide solution. The unit cell has been determined by single crystal X-ray diffraction but the crystal structure was not reported. An infrared spectroscopic analysis indicates that the charge transfer between the BEDT-TTF and C_{60} molecules is quite small. It has been suggested that such complexes may offer a means to introduce iron into the fullerene lattice.

9 Platinum and Mercury Organometallic Derivatives of TTF

TTF and 3,4-dimethyltetrathiafulvalene (*o*-Me_2TTF) react with $[Pt(\eta^2\text{-}C_2H_4)(PPh_3)_2]$ to form the organometallic $[Pt(\eta^2\text{-TTF})(PPh_3)_2]$ and $[Pt(\eta^2\text{-}o\text{-}Me_2TTF)(PPh_3)_2]$ complexes through ethylene displacement [105] (see Scheme 8). The crystal structure of the later of these confirms that TTF acts as a π acidic ligand: the molecular structure is characterized by a trigonal arrangement of the alkene in the plane of the triangle.

One example of a –Hg– linked TTF dimer has been reported [106]. Linear coordination is expected for the mercury atom. Extended Hückel calculations indicate that the mercury orbitals are essentially not involved in the HOMOs, and thus the two TTF redox groups can be regarded as being completely isolated from each other. The voltammograms confirm these expectations in that both TTF

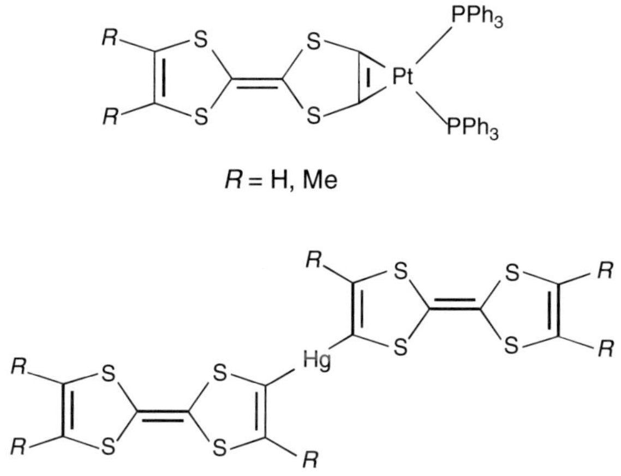

R = H, Me

Scheme 8 Platinum and mercury containing organometallic derivatives of TTF

fragments are oxidized at the same potential. TCNQ salts of these mercury dimers have been synthesized by chemical oxidation. The compressed pellet conductivity of these polycrystalline samples indicates semiconductive behavior with a conductivity of $\sim 10^{-2}$ S cm^{-1} and an activation energy of 0.2 eV. Electrocrystallization has been difficult because of the low solubility of the TTF mercury dimer. However, microcrystals of the $(BF_4)^-$, $(ClO_4)^-$, $(PF_6)^-$, and SCN^- salts have been prepared, but further characterization has not been reported.

10 Conclusions

The field of molecular conductors began more than half a century ago with the discovery of high conductivity in a perylene bromide complex [107]. Metallic-like conductivity was reported for the first time in the TTF-TCNQ charge transfer salt about 20 years later [2, 3] with the first superconducting organic material, $(TMTSF)_2PF_6$, discovered in 1979 [4]. Since then, the field has expanded rapidly with hundreds of cation-radical salts being reported in the literature with a wide assortment of anions. When one considers the vast assortment of available organometallic anions, it is surprising that only a handful have been used as components of these salts. The most studied of these salts belong to the κ-$(BEDT\text{-}TTF)_2M(CF_3)_4$(1,1,2-trihaloethane) (M = Cu, Ag and Au) family as these form one of the most diverse sets of molecular superconductors known. A handful of anions related to the $[M(CF_3)_4]^-$ family have also been used as components of cation-radical salts. A number of halophenylene-based anions have also been studied, but their structural and property characterizations are less complete.

Interest in the incorporation of organometallic anions into molecular conductors has increased over the past decade as attention has been given to the design of hybrid multifunctional materials that contain conducting and magnetic moieties in the same crystal lattice. Metallocenium anions have been especially studied in this regard, both as discrete anions and as entities covalently tethered to TTF molecules. Many organometallic anions possess open coordination sites that allow for direct coordination to chalcogenide-based electron donor molecules.

The use of organometallic anions as components of conductive salts appears to be poised for significant growth in the years ahead. The "molecular toolbox" is well stocked with these molecular building blocks and their ability to form various types of bonds (covalent, coordination, hydrogen, etc.) with electron donor molecules promises to make this a rich area for future study. Although many claims have been made to "rationally" design new materials, computational chemistry is not yet able to make reliable predictions of structures based on the molecular components. While systematic advances are continually being made in the field, we eagerly await the occasional scientific breakthroughs that open new vistas for materials research.

Acknowledgments Work supported by UChicago Argonne, LLC, Operator of Argonne National Laboratory ("Argonne"). Argonne, a U.S. Department of Energy Office of Science laboratory, is operated under Contract No. DE-AC02-06CH11357.

References

1. Wudl F, Smith GM, Hufnagel EJ (1970) J Chem Soc Chem Commun 1453
2. Coleman LB, Cohen MJ, Sandman DJ, Yamagishi FG, Garito AF, Heeger AJ (1973) Solid State Commun 12:1125
3. Ferraris J, Cowan DO, Walatka VJ, Perlstein JH (1973) J Am Chem Soc 95:948
4. Jérome D, Mazaud A, Ribault M, Bechgaard K (1980) J Phys Lett 41:L95
5. Bechgaard K, Carneiro K, Olsen M, Rasmussen FB, Jacobsen CS (1981) Phys Rev Lett 46:852
6. Mori T (1998) Bull Chem Soc Jpn 71:2509
7. Mori T, Mori H, Tanaka S (1999) Bull Chem Soc Jpn 72:179
8. Mori T (1999) Bull Chem Soc Jpn 72:2011
9. Urayama H, Yamochi H, Saito G, Nozawa K, Sugano T, Kinoshita M, Sato S, Oshima K, Kawamoto A, Tanaka J (1988) Chem Lett 55
10. Kini AM, Geiser U, Wang HH, Carlson KD, Williams JM, Kwok WK, Vandervoort KG, Thompson JE, Stupka DL, Jung D, Whangbo M-H (1990) Inorg Chem 29:2555
11. Williams JM, Kini AM, Wang HH, Carlson KD, Geiser U, Montgomery LK, Pyrka GJ, Watkins DM, Kommers JM, Boryschuk SJ, Strieby Crouch AV, Kwok WK, Schirber JE, Overmyer DL, Jung D, Whangbo M-H (1990) Inorg Chem 29:3272
12. Geiser U, Schlueter JA (2004) Chem Rev 104:5203

13. Williams JM, Ferraro JR, Thorn RJ, Carlson KD, Geiser U, Wang HH, Kini AM, Whangbo MH (1992) Organic superconductors (including fullerenes). Prentice Hall, Englewood Cliffs, New Jersey
14. Ishiguro T, Yamaji K, Saito G (1998) Organic superconductors, vol 88, 2nd edn. Springer, Berlin Heidelberg New York
15. Toyota N, Lang M, Muller J (2007) Low-dimensional molecular metals. Springer, Berlin Heidelberg New York
16. Batail P (2004) Chem Rev 104
17. Brooks JS (2007) In: Schrieffer JR, Brooks JS (eds) Handbook of high-temperature superconductivity. Springer, Berlin Heidelberg New York, p 463
18. Wosnitza J (2007) J Low Temp Phys 146:641
19. Singleton J (2000) Rep Prog Phys 63:1111
20. Powell BJ, McKenzie RH (2006) J Phys Condens Matter 18:R827
21. Singleton J, Mielke C (2002) Contemp Phys 43:63
22. Dukat W, Naumann D (1986) Rev Chim Miner 23:589
23. Naumann D, Roy T, Tebbe K-F, Crump W (1993) Angew Chem Int Ed Engl 32:1482
24. Geiser U, Schlueter JA, Wang HH, Williams JM, Nauman D, Roy T (1995) Acta Crystallogr B 51:789
25. Emge TJ, Wang HH, Beno MA, Leung PCW, Firestone MA, Jenkins HC, Cook JD, Carlson KD, Williams JM, Venturini EL, Azevedo LJ, Schirber JE (1985) Inorg Chem 24:1736
26. Schlueter JA, Geiser U, Williams JM, Wang HH, Kwok WK, Fendrich JA, Carlson KD, Achenbach CA, Dudek JD, Naumann D, Roy T, Schirber JE, Bayless WR (1994) J Chem Soc Chem Commun 1599
27. Schlueter JA, Carlson KD, Geiser U, Wang HH, Williams JM, Kwok WK, Fendrich JA, Welp U, Keane PM, Dudek JD, Komosa AS, Naumann D, Roy T, Schirber JE, Bayless WR, Dodrill B (1994) Physica C 233:379
28. Schlueter JA, Williams JM, Geiser U, Dudek JD, Sirchio SA, Kelly ME, Gregar JS, Kwok WH, Fendrich JA, Schirber JE, Bayless WR, Naumann D, Roy T (1995) J Chem Soc Chem Commun 1311
29. Schlueter JA, Geiser U, Kini AM, Wang HH, Williams JM, Naumann D, Roy T, Hoge B, Eujen R (1999) Coord Chem Rev 190/192:781
30. Schlueter JA, Williams JM, Geiser U, Dudek JD, Kelly ME, Sirchio SA, Carlson KD, Naumann D, Roy T, Campana CF (1995) Adv Mater 7:634
31. Schlueter JA, Geiser U, Wang HH, Kelly ME, Dudek JD, Williams JM, Naumann D, Roy T (1996) Mol Cryst Liq Cryst 284:195
32. Müller J, Lang M, Steglich F, Schlueter JA, Kini AM, Geiser U, Mohtasham J, Winter RW, Gard GL, Sasaki T, Toyota N (2000) Phys Rev B 61:11739
33. Schlueter JA, Geiser U, Williams JM, Dudek JD, Kelly ME, Flynn JP, Wilson RR, Zakowicz HI, Sche PP, Naumann D, Roy T, Nixon PG, Winter RW, Gard GL (1997) Synth Met 85:1453
34. Kini AM, Schlueter JA, Ward BH, Geiser U, Wang HH (2001) Synth Met 120:713
35. Schlueter JA, Williams JM, Kini AM, Geiser U, Dudek JD, Kelly ME, Flynn JP, Naumann D, Roy T (1996) Physica C 265:163
36. Schlueter JA, Williams JM, Geiser U, Wang HH, Kini AM, Kelly ME, Dudek JD, Naumann D, Roy T (1996) Mol Cryst Liq Cryst 285:43
37. Schlueter JA, Kini AM, Ward BH, Geiser U, Wang HH, Mohtasham J, Winter RW, Gard GL (2001) Physica C 351:261
38. Kini AM, Wang HH, Schlueter JA, Dudek JD, Sirchio SA, Carlson KD, Williams JM (1996) Physica C 264:81
39. Geiser U, Schlueter JA, Carlson KD, Williams JM, Wang HH, Kwok WK, Welp U, Fendrich JA, Dudek JD, Achenbach CA, Komosa AS, Keane PM, Naumann D, Roy T, Schirber JE, Bayless WR, Ren J, Whangbo MH (1995) Synth Met 70:1105

40. Eldridge JE, Xie Y, Schlueter JA, Williams JM, Naumann D, Roy T (1996) Solid State Commun 99:335
41. Geiser U, Schultz AJ, Wang HH, Watkins DM, Stupka DL, Williams JM, Schirber JE, Overmyer DL, Jung D, Novoa JJ, Whangbo M-H (1991) Physica C 174:475
42. Eldridge JE, Kornelsen K, Wang HH, Williams JM, Strieby Crouch AV, Watkins DM (1991) Solid State Commun 79:583
43. Wosnitza J, Beckmann D, Wanka S, Schlueter JA, Williams JM, Naumann D, Roy T (1996) Solid State Commun 98:21
44. Geiser U, Schlueter JA, Williams JM, Kini AM, Dudek JD, Kelly ME, Naumann D, Roy T (1997) Synth Met 85:1465
45. Wang HH, VanZile ML, Geiser U, Schlueter JA, Williams JM, Kini AM, Sche PP, Nixon PG, Winter RW, Gard GL, Naumann D, Roy T (1997) Synth Met 85:1533
46. Schlueter JA, Carlson KD, Williams JM, Wang HH, Geiser U, Welp U, Kwok WK, Fendrich JA, Dudek JD, Achenbach CA, Keane PM, Komosa AS, Naumann D, Roy T, Schirber JE, Bayless WR (1994) Physica C 230:378
47. Schirber JE, Venturini EL, Kini AM, Wang HH, Whitworth JR, Williams JM (1988) Physica C 152:157
48. Geiser U, Schlueter JA, Dudek JD, Williams JM, Naumann D, Roy T (1995) Acta Crystallogr C 51:1779
49. Wang HH, Schlueter JA, Geiser U, Williams JM, Naumann D, Roy T (1995) Inorg Chem 34:5552
50. Eujen R, Hoge B, Brauer DJ (1997) Inorg Chem 36:1464
51. Eujen R, Hoge B, Brauer DJ (1997) Inorg Chem 36:3160
52. Cerrada E, Garin J, Gimeno MC, Laguna A, Laguan M, Jones PG (1993) Special Publication R Soc Chem 131:164
53. Cerrada E, Laguna M, Bartolome J, Campo J, Orera V, Jones PG (1998) Synth Met 92:245
54. Usón R, Laguna A, Vicente J, Garcia J, Jones PG, Sheldrick GM (1981) J Chem Soc Dalton Trans 655
55. Schlueter JA, Geiser U, Wang HH, Van Zile ML, Fox SB, Williams JM, Laguna A, Laguna M, Naumann D, Roy T (1997) Inorg Chem 36:4265
56. Haneline MR, Gabbai FP (2004) C R Chimie 7:871
57. Miller JS, Calabrese JC, Rommelmann H, Chittipeddi SR, Zhang JH, Reiff WM, Epstein AJ (1987) J Am Chem Soc 109:769
58. Broderick WE, Thompson JA, Day EP, Hoffmann BM (1990) Science 249:401
59. Forward JM, Mingos DMP, Muller TE, Williams DJ, Yan Y-K (1994) J Organomet Chem 467:207
60. Marsh RE, Spek AL (2001) Acta Cryst B 57:800
61. Yan Y-K, Mingos DMP, Kurmoo M, Li W-S, Scowen IJ, McPartlin M, Coomber AT, Friend RH (1995) J Chem Soc Chem Commun 997
62. Yan Y-K, Mingos DMP, Kurmoo M, Li W-S, Scowen IJ, McPartlin M, Coomber AT, Friend RH (1995) J Chem Soc Dalton Trans 2851
63. Yan Y-K, Mingos DMP, Williams DJ, Kurmoo M (1995) J Chem Soc Dalton Trans 3221
64. Guionneau P, Kepert CJ, Bravic G, Chasseau D, Truter MR, Kurmoo M, Day P (1997) Synth Met 86:1973
65. Miller JS, Epstein AJ, Reiff WM (1988) Chem Rev 88:201
66. Morse DB, Rauchfuss TB, Wilson SR (1988) J Am Chem Soc 110:2646
67. Bottomley F, Darkwa J, Sutin L, White PS (1986) Organometallics 5:2165
68. Furuta K, Akutsu H, Yamada J-i, Nakatsuji Si (2004) Chem Lett 33:1214
69. Furuta K, Akutsu H, Yamada J-i, Nakatsuji Si, Turner SS (2006) J Mater Chem 16:1504
70. Akutsu H, Ohnishi R, Yamada J-i, Nakatsuji Si, Turner SS (2007) Inorg Chem 46:8472
71. Furuta K, Akutsu H, Yamada J-i, Nakatsuji Si (2004) Synth Met 152:381
72. Sarhan AE-WAO (2005) Tetrahedron 61:3889
73. Ueno Y, Sano H, Okawara M (1980) J Chem Soc Chem Commun 28

74. Bryce MR, Skabara PJ, Moore AJ, Batsanov AS, Howard JAK, Hoy VJ (1997) Tetrahedron 53:17781
75. Iyoda M, Takano T, Otani N, Ugawa K, Yoshida M, Matsuyama H, Kuwatani Y (2001) Chem Lett 1310
76. Lee H-J, Noh D-Y, Underhill AE, Lee C-S (1999) J Mater Chem 9:2359
77. Moore AJ, Skabara PJ, Bryce MR, Batsanov AS, Howard JAK, Daley STAK (1993) J Chem Soc Chem Commun 417
78. Moore AJ, Bryce MR, Skabara PJ, Batsanov AS, Goldenberg LM, Howard JAK (1997) J Chem Soc Perkin Trans 1:3443
79. Wilkes SB, Butler IR, Underhill AE, Kobayashi A, Kobayashi H (1994) J Chem Soc Chem Commun 53
80. Wilkes SB, Butler IR, Underhill AE, Hursthouse MB, Hibbs DE, Malik KMA (1995) J Chem Soc Dalton Trans 897
81. Kawabata K, Makano M, Tamura H, Matsubayashi G-e (2004) Eur J Inorg Chem 2137
82. Kawabata K, Nakano M, Tamura H, Matsubayashi G-e (2005) Inorg Chim Acta 358:2082
83. Mori H, Nakano M, Tamura H, Matsubayashi G-e (1999) J Organomet Chem 574:77
84. Saito K, Nakano M, Tamura H, Matsubayashi G-e (2000) Inorg Chem 39:4815
85. Saito K, Nakano M, Tamura H, Matsubayashi G-e (2001) J Organomet Chem 625:7
86. McCullough RD, Belot JA, Rheingold AL, Yap GPA (1995) J Am Chem Soc 117:9913
87. McCullough RD, Belot JA, Seth J, Rheingold AL, Yap GPA, Cowan DO (1995) J Mater Chem 5:1581
88. Fourmigue M, Batail P (1992) Bull Soc Chim Fr 129:29
89. Fourmigue M, Uzelmeier CE, Boubekeur K, Bartley SL, Dunbar KR (1997) J Organomet Chem 529:343
90. Avarvari N, Martin D, Fourmigue M (2002) J Organomet Chem 643/644:292
91. Gouverd C, Biaso F, Cataldo L, Berclaz T, Geoffroy M, Levillain E, Avarvari N, Fourmigue M, Sauvage F, Wartelle C (2005) Phys Chem Chem Phys 7:85
92. Avarvari N, Fourmigue M (2004) Chem Commun 1300
93. Pellon P, Gachot G, Le Bris J, Marchin S, Carlier R, Lorcy D Inorg Chem 42:2056
94. Matsubayashi G-e, Ryowa T, Tamura H, Nakano M, Arakawa R (2002) J Organomet Chem 645:94
95. Matsubayashi G-e, Yokoyama K, Tanaka T (1998) J Chem Soc Dalton Trans 253
96. Vallarino LM (1965) Inorg Chem 4:161
97. Kubo K, Nakano M, Tamura H, Matsubayashi G-e, Nakamoto M (2003) J Organomet Chem 669:141
98. Kubo K, Nakao A, Ishii Y, Kato R, Matsubayashi G-e (2005) Synth Met 153:425
99. Kubo K, Nakano M, Tamura H, Matsubayashi G-e (2003) Eur J Inorg Chem 4093
100. Horiuchi S, Yamochi H, Saito G, Sakaguchi K-i, Kusunoki M (1996) J Am Chem Soc 118:8604
101. Kubo K, Nakao A, Ishii Y, Tamura M, Kato R, Matsubayashi G-e (2006) J Low Temp Phys 142:413
102. Kubo K, Nakao A, Ishii Y, Yamamoto T, Tamura M, Kato R, Yakushi K, Matsubayashi G-e (2008) Inorg Chem 47:5495
103. Suga Y, Nakano M, Tamura H, Matsubayashi G-e (2004) Bull Chem Soc Jpn 77:1877
104. Spitsina NG, Semkin VN, Graja A, Pukacki W, Dyachenko OA, Gritsenko VV (1996) Polish J Chem 70:70
105. Jayaswal MN, Peindy HN, Guyon F, Knorr M, Ararvari N, Fourmigue M (2004) Eur J Inorg Chem:2646
106. Fourmigue M, Huan Y-S (1993) Organometallics 12:797
107. Akamatu H, Inokuchi H, Matsunaga Y (1954) Nature 173:168

Top Organomet Chem (2009) 27: 35–53

New Molecular Architecture for Electrically Conducting Materials Based on Unsymmetrical Organometallic-Dithiolene Complexes

Kazuya Kubo and Reizo Kato

Abstract New molecular architecture for highly conducting molecular materials was developed with use of unsymmetrical organometallic-dithiolene complexes. The new architecture has various advantages including easy modification of their molecular and electronic features. Organometallic complexes based on unsymmetrical Au(III)-dithiolene complexes [(ppy)Au($C_8H_4S_8$ or $C_8H_4S_6O_2$)] were prepared for new cationic components of molecular conductors. These unsymmetrical organometallic complexes can provide various cation radical salts $[(ppy)Au(S\text{-}S)]_2[anion][solvent]_n$ (S-S = $C_8H_4S_8$ or $C_8H_4S_6O_2$, anion = PF_6^-, BF_4^-, AsF_6^-, TaF_6^-, solvent = PhCl, n = 0–0.5) by constant current electrolysis of their benzonitrile or chlorobenzene solutions containing (Bu_4N)(anion) as electrolyte. $[(ppy)Au(C_8H_4S_8)]_2[PF_6]$ under pressure is the first molecular metal based on the organometallic component. In this review, principle of the molecular architecture based on the unsymmetrical organometallic-dithiolene complexes and physical properties of their cation radical salts are discussed.

Keywords Donor type metal complexes, Metal dithiolene complexes, Molecular conductors, Organometallic complexes, Unsymmetrical metal complexes

Contents

K. Kubo(✉) and R. Kato
Condensed Molecular Materials Laboratory, 2-1 Hirosawa, Wako-shi, Saitama, 351-0198, Japan,
E-mail: kkubo@riken.jp

M. Fourmigué and L. Ouahab (eds.), *Conducting and Magnetic Organometallic Molecular Materials*, Topics in Organometallic Chemistry 27,
DOI: 10.1007/978-3-642-00408-7_2,

Abbreviations

BEDT-TTF	Bis(ethylenedithio)tetrathiafulvalene
bpy	2,2′-Bipyridine
Bu-pia	*N*-Butyl-pyridine-carbaldimine
CH_2Cl_2	Dichloromethane
$C_8H_4S_8{}^{2-}$	2-{(4,5-Ethylenedithio)-1,3-dithiole-2-ylidene}-1,3-dithiole-4,5-dithiolate(2-)
$C_8H_4S_6O_2{}^{2-}$	2-{(4,5-Ethylenedioxio)-1,3-dithiole-2-ylidene}-1,3-dithiole-4,5-dithiolate(2-)
CT	Charge transfer
DCNQI	*N,N′*-Dicyanoquinonediimine
dddt	5,6-Dihydro-1,4-dithiin-2,3-dithiolate
Dec-pia	*N*-Decyl-pyridine-carbaldimine
$dmit^{2-}$	1, 3-Dithiole-2-thiol-4,5-dithiolate(2-)
Et-pia	*N*-Ethyl-pyridine-carbaldimine
Hexdec-pia	*N*-Hexadecyl-pyridine-carbaldimine
HOMO	Highest occupied molecular orbital
IR	Infrared
LUMO	Lowest unoccupied molecular orbital
mnt	Maleonitriledithiolate
MO	Molecular orbital
PhCl	Chlorobenzene
phen	1,10-Phenanthoroline
ppy^-	*C*-Dehydro-2-phenylpridine
Pr^i-pia	*N*-Iso-propyl-pyridine-carbaldimine
TCNQ	7,7,8,8-Tetracyano-*p*-quinodimethane
tmdt	Trimethylenetetrathiafulvalenedithiolate
TTF	Tetrathiafulvalene

1 General Introduction of Molecular Conductors

1.1 General Features of Molecular Conductors

Since the discovery of the first organic semiconductor perylene-bromine complex in 1954 [1], a large number of molecular conductors, including more than 100 molecular superconductors, have been prepared. Conducting molecular materials are characterized by the following features;

1. Clear and simple electronic structure which can be well described by the tight-binding band picture based on the extended Hückel molecular orbital calculation [2]
2. A variety of physical properties that originate from low-dimensionality, strongly correlated electron-electron interaction, frustration effect, and so on
3. Softness of the crystal lattice and sensitivity to external stimuli
4. Plenty of possibilities for chemical design

Thanks to these characteristics, the molecular conductor is now one of the standard materials in condensed matter physics and interdisciplinary basic science is expanding.

In general, component molecules for molecular conductors (Fig. 1) belong to the π-conjugated system and are divided into two categories, electron donor and electron acceptor. The component molecule is an insulator in itself. In order to obtain the metallic state, the formation of at least one partially filled energy band is required. A straightforward access to the molecular metal can be achieved by arranging open-shell molecules (radicals) so as to enable intermolecular electron transfer. In most cases, cation radicals or anion radicals generated from the donor or the acceptor have been used for the formation of metallic molecular crystals, and the conduction band originate from HOMO of the donor or LUMO of the acceptor. It should be added that electron transfer between HOMO and LUMO bands in a single-component molecular crystal can also generate partially filled energy bands, which is observed in the first single-component molecular metal $Ni(tmdt)_2$ (Fig. 1) where an energy gap between HOMO and LUMO is very small and intermolecular interactions are sufficiently large [3].

Since planar π-conjugated molecules tend to stack to form the column structure, molecular metals developed in the early stage have the one-dimensional electronic structure which is characterized by a pair of planar Fermi surfaces and is associated with a metal-insulator transition at low temperatures. In order to achieve the stable metallic state down to low temperatures, much effort to increase the dimensionality of the electronic structure has been made by means of chemical modification and/or application of pressure. An organic donor BEDT-TTF (Fig. 1), where sulfur-containing six-membered rings attached to the TTF moiety effectively enhance the two-dimensional intermolecular interaction, opened a new molecular design for higher dimensional electronic structure and provided various types of metallic and superconducting cation radical salts [4]. In the (2,5-disubstituted DCNQI)-Cu

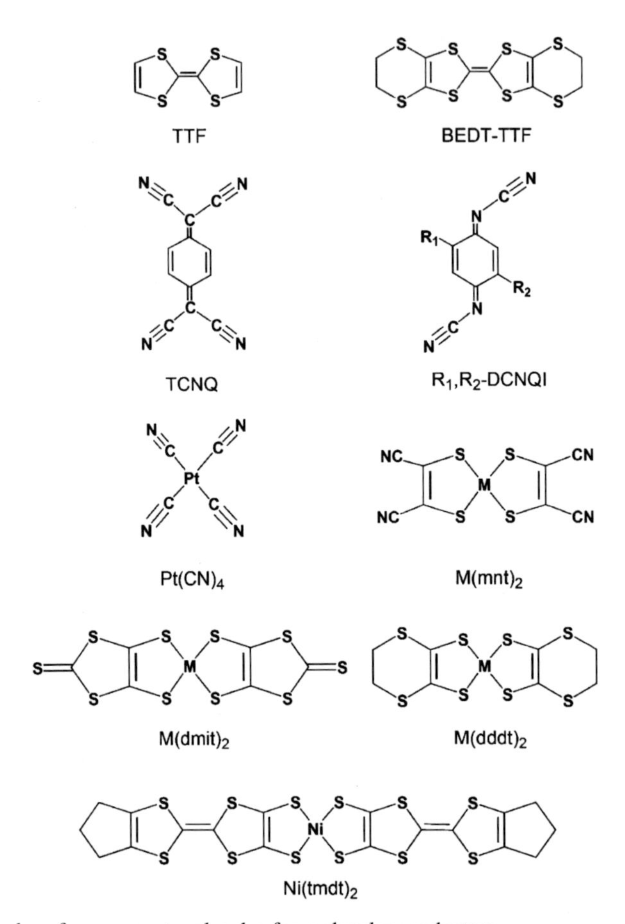

Fig. 1 Examples of component molecules for molecular conductors

salts, the *pπ-d* interaction between the acceptor molecule and the tetrahedrally coordinated Cu ion (in the mixed-valence state) forms a three-dimensional electronic structure [5].

Various mechanisms make the metallic state in the molecular system unstable. One is density wave (charge density wave and spin density wave) formation in the one-dimensional system, which can be removed by the increase of the dimensionality of the electronic structure. Even in the higher dimensional systems, however, the strong electron-electron correlation can induce the nonmetallic state. In systems with a half-filled energy band, where each unit in the crystal has one conduction electron, strong Coulomb interaction within the same site can induce an insulating state if the conduction band is narrow. This is a Mott insulating state. On the other hand, (typically) in systems with a quarter-filled band, strong Coulomb interaction between neighboring sites can induce inhomogeneous distribution of the site charges. This is a charge ordered insulating state.

The formal charge of +1/2 (−1/2) is the frequently observed valence state in the donor (acceptor) molecule in molecular conductors, which provides a quarter-filled hole (electron) band where the charge ordered state can occur. When the molecules are dimerized, the conduction band splits into bonding and antibonding pairs. If the degree of dimerization is strong, the upper and lower bands are largely separated by the dimerization gap and one of them is effectively half-filled. In this case, the system can behave like a half-filled system (one conduction electron on each dimer) and exhibit the Mott insulating state.

1.2 Molecular Conductors Based on Metal Complexes

In the early stage of the development of molecular conductors based on metal complexes, partially oxidized tetracyanoplatinate salts (for example, KCP; $K_2[Pt(CN)_4]Br_{0.30}\cdot 3H_2O$) and related materials were intensively studied [6]. In this system, the square-planar platinum complexes are stacked to form a linear Pt-atom chain. The conduction band originates from the overlap of $5dz^2$ orbitals of the central platinum atom and exhibits the one-dimensional character.

On the other hand, metal dithiolene complexes possess a delocalized electron system as a planar central core $M(C_2S_2)_2$. The conduction band is formed by the ligand π orbitals or mixed-metal-ligand orbitals where the sulfur atoms play an important role [7]. Depending on the choice of substituent groups attached to the central core, metal dithiolene complexes behave as both the donor and the acceptor. Development of molecular conductors based on the dithiolene complexes was triggered by the discovery of the metallic behavior in an anion radical salt $(H_3O)_{0.33}Li_{0.8}[Pt(mnt)_2]\cdot 1.67H_2O$ [8]. Among metal dithiolene complexes, the metal-dmit complexes $M(dmit)_2$ (M = Ni and Pd; Fig. 1) have been the most studied system. In the $M(dmit)_2$ molecule, HOMO has b_{1u} symmetry, while LUMO has b_{2g} symmetry. The metal d orbital can mix into the LUMO, but cannot contribute to the HOMO due to the symmetry, which destabilizes the HOMO and leads to a small energy splitting between HOMO and LUMO. The side-by-side intermolecular interaction, which leads to the formation of the two-dimensional electronic structure, is strong for the HOMO and weak for the LUMO. This is because some of overlap integrals for the intermolecular S···S pairs are canceled out due to the b_{2g} symmetry of the LUMO. Although the $M(dmit)_2$ molecule belongs to the acceptor, the nature of the conduction band in their anion radical salts strongly depends on the central metal. In general, the conduction band of the Pd system originates from the HOMO, while the conduction band of the Ni system originates from the LUMO. This unusual feature of the Pd system, HOMO-LUMO band inversion, is due to the strong dimerization and the small energy splitting between HOMO and LUMO [9]. In a series of anion radical salts with closed-shell cations (Cation)$[Pd(dmit)_2]_2$ (Cation = $Et_xMe_{4-x}Z^+$; Z = N, P, As, Sb and $x = 0, 1, 2$), the dimer units $[Pd(dmit)_2]_2^-$ form a strongly correlated two-dimensional system with a quasi triangular lattice and exotic properties derived from frustration and strong correlation are reported [7]. In these Pd salts, the choice of the counter cation tunes the degrees of frustration and correlation which are associated with the molecular arrangement. On the other hand, in the Ni salts, the choice of the counter cation provides a variety of molecular arrangements [7].

1.3 Molecular Conductors Based on Cationic Metal Complexes

Compared with the conducting anion radical salts of metal complexes, the number of molecular conductors based on cationic metal complexes is still limited. Donor type complexes $M(dddt)_2$ (M = Ni, Pd, Pt; Fig. 1) are the most studied system. The $M(dddt)_2$ molecule is a metal complex analogue of the organic donor BEDTTTF. Formally, the central C=C bond of BEDT-TTF is substituted by a metal ion. The HOMO and LUMO of the $M(dddt)_2$ molecule are very similar in orbital character to those of the $M(dmit)_2$ molecule. In addition, the HOMO of the $M(dddt)_2$ molecule is also very similar to that of BEDT-TTF. More than ten cation radical salts of $M(dddt)_2$ with a cation:(monovalent) anion ratio of 2:1 or 3:2 are reported [7]. A few of them exhibit metallic behavior down to low temperatures. The HOMO-LUMO band inversion can also occur in the donor system depending on the degree of dimerization. In contrast to the acceptor system, however, the HOMO-LUMO band inversion in the donor system leads a LUMO band with the one-dimensional character to the conduction band.

2 Unsymmetrical Components Based on Diimine–Dithiolene Complexes

2.1 Molecular Design of Unsymmetrical Components Based on Diimine–Dithiolene Complexes

Although there are a large number of symmetrical metal-dithiolene complexes as described in the previous section [7], only a few applications of mixed-ligand complexes have been reported as conducting materials [10]. Among various mixed-ligand complexes, a donor (HOMO)-metal-acceptor (LUMO) type molecule is capable of providing a new principle for the design of molecular conductors utilizing a potential interplay of intra- and intermolecular charge transfers (Fig. 2).

Pioneering work in this field was started in the 1980s by Matsubayashi et al. They suggested that planar $[(N\text{-}N)M(S\text{-}S)]^n$ (M = Ni^{2+}, Pd^{2+}, Pt^{2+}, Au^{3+}; N-N = phen, bpy and Et-, Pr^i-, Bu-, Dec-, Hexdec-pia; $S\text{-}S^{2-}$ = $dmit^{2-}$ and $C_8H_4S_8{}^{2-}$; $n = 0$ or +1) type unsymmetrical metal-dithiolene complexes would be good candidates for new components of molecular conductors [11–14]. The complexes, having HOMO on the dithiolene ligand and LUMO on the diimine ligand, are known to exhibit remarkable emission and luminescent spectra, due to their

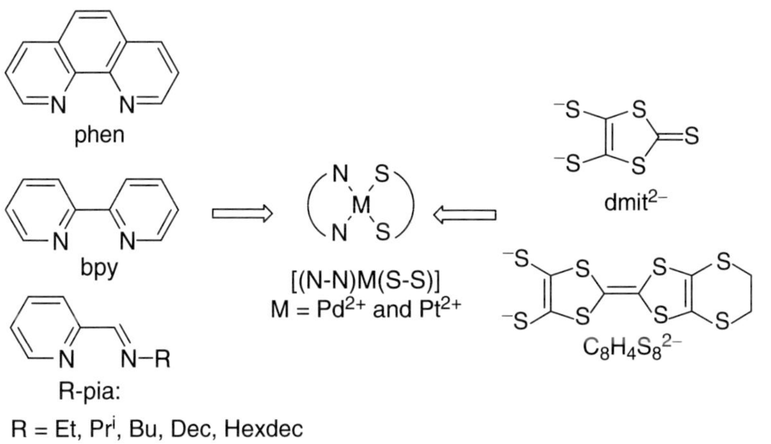

Fig. 2 Schematic drawing of unsymmetrical donors based on diimine-dithiolate complexes

unconventional electronic structures [15–21]. They have attempted to expand the field of conducting and magnetic materials by using the unsymmetrical molecules. For this purpose, a series of the diimine Pd^{2+} and Pt^{2+} complexes with the dithiolene ligands were prepared.

2.2 *Cation Radical Salts of the Unsymmetrical Diimine–Dithiolene Complexes*

2.2.1 Electrical Resistivities of the Cation Radical Salts

Matsubayashi et al. revealed donor abilities of the unsymmetrical diimine–dithiolene complexes [11–14]. The unsymmetrical complexes provided cation radical salts with various anions including I_3^-, Br_3^- and $TCNQ^-$ by use of chemical oxidation [11–14]. The electrical resistivities of the cation radical salts measured with their compressed pellets at room temperature are summarized in Table 1. The electrical resistivities of the dmit complexes were very high. The cation radical salts of the $C_8H_4S_8$-complexes, which have the BEDT-TTF moiety [22, 23], exhibited lower resistivity than those of dmit complexes, except for [(Bu-pia)Pt($C_8H_4S_8$)] salts. However, crystal structures of these salts were not reported, and details of their electrical properties and electronic states were not discussed based on their crystal structures.

2.2.2 Crystal Structure of the Cation Radical Salt [(bpy)Pt($C_8H_4S_8$)][BF_4]

Crystal structure data are indispensable for the discussion of the conduction mechanism in the cation radical salts based on the unsymmetrical complexes. In 2002,

Table 1 Cation radical salts based on the unsymmetrical donors and their electrical resistivities at room temperature

Complex	$\rho_{r.t.}$ Ω^{-1} cm^{-1}	Ref.
$[(phen)Pt(dmit)]I_{2.2}$	$>10^5$	[4]
$[(bpy)Pt(dmit)]I_{1.9}$	$>10^5$	[4]
$[(Et\text{-}pia)Pt(dmit)]I_{1.9}$	$>10^5$	[4]
$[(Et\text{-}pia)Pt(dmit)]I_{3.4}$	$>10^5$	[3]
$[(Et\text{-}pia)Pt(dmit)]Br_{3.1}$	$>10^5$	[3]
$[(Pr^i\text{-}pia)Pt(dmit)]I_{1.7}$	$>10^5$	[4]
$[(Dec\text{-}pia)Pt(dmit)]I_{3.3}$	$>10^5$	[5]
$[(Hexdec\text{-}pia)Pt(dmit)]I_{2.1}$	$>10^5$	[5]
$[(Bu\text{-}pia)Pd(dmit)]I_{1.8}$	$>10^5$	[6]
$[(bpy)Pt(C_8H_4S_8)]I_{2.7}$	7.7×10^2	[6]
$[(bpy)Pt(C_8H_4S_8)]TCNQ_{0.8}$	1.5×10^2	[6]
$[(Bu\text{-}pia)Pt(C_8H_4S_8)]I_3$	$>10^5$	[6]
$[(Bu\text{-}pia)Pt(C_8H_4S_8)]TCNQ_{0.6}$	$>10^5$	[6]
$[(Bu\text{-}pia)Pd(C_8H_4S_8)]I_{5.1}$	7.1×10^2	[6]
$[(Bu\text{-}pia)Pd(C_8H_4S_8)]TCNQ_{0.4}$	3.1×10^2	[6]

the first crystal structure analysis of a 1:1 cation radical salt based on the unsymmetrical diimine-dithiolene Pt complex $[(bpy)Pt(C_8H_4S_8)][BF_4]$ was reported by Kubo et al. [24]. Constant current electrolysis of a CH_2Cl_2 solution of $[(bpy)Pt(C_8H_4S_8)]$ in the presence of $(Bu_4N)(BF_4)$ afforded needles of a one-electron oxidized species $[(bpy)Pt(C_8H_4S_8)][BF_4]$ [14, 24, 25]. The crystal structure viewed along the *b* and *c* axes is shown in Fig. 3. The cation moieties are stacked with the same molecular orientation to form columnar structure along the *c* axis. S···S contacts shorter than van der Waals radii (<3.7 Å) are observed within and between the columns. The electrical resistivity of the salt measured with a compressed pellet at room temperature was 2.5×10^2 Ω cm. However, no data of temperature dependence of resistivity and electronic structure has been reported.

2.3 Conclusion

Matsubayashi et al. developed a new research field of molecular conductors by using the unsymmetrical diimine-dithiolene complexes. Various components with dmit and $C_8H_4S_8$ as dithiolene ligands were prepared. The unsymmetrical metal complexes provided various cation radical salts. Their electrical resistivities were measured with their compressed pellets at room temperature. The salt of the complexes with the $C_8H_4S_8$ ligand exhibited lower resistivity than that of the salt with the dmit ligand. No crystallographic data was reported. Thus, the details of the conducting mechanism of these salts cannot be discussed. The first crystal structure analysis was reported for the 1:1 salt $[(bpy)Pt(C_8H_4S_8)][BF_4]$.

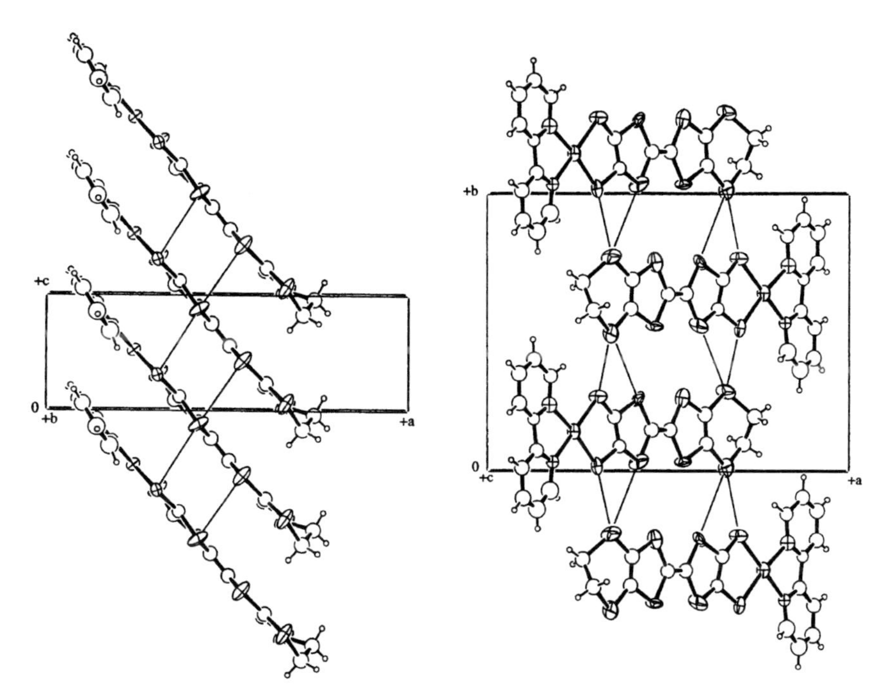

Fig. 3 Crystal structure depicted on the basis of the result described in [24]. (Reprinted with permission from [24]. Copyright 2002 Elsevier)

3 Unsymmetrical Components Based on Organometallic–Dithiolene Complexes

3.1 General Features of Unsymmetrical Components Based on Organometallic Metal–Dithiolene Complexes

The electronic state of this type of mixed-ligand unsymmetrical components can be modified by the combination of metal ions, nitrogen and sulfur-ligands. Kubo et al. modified the unsymmetrical complexes by an introduction of the carbon-metal σ-bond to improve the crystal quality and the conducting properties of their cation radical salts. They prepared the organometallic donor [(ppy)Au($C_8H_4S_8$ or $C_8H_4S_8O_6$)] (Scheme 1) and studied the electronic structures of the molecules [26, 27]. The organometallic gold complexes have the same molecular structure and charge with those of [(bpy)Pt($C_8H_4S_8$)] (see Sect. 2.1). However, orthometalated chelating ligands such as ppy^- can modify the electronic structure of diimine complexes, because of its asymmetry and the strong σ-bonding of the phenyl carbon atom [28–34]. The distributions of the HOMO and LUMO in these orthometalated compounds are analogous to those of the diimine complexes (Fig. 4) [36]. However, the lowest charge transfer excited states of the square-planar d_8 complex [(ppy)Au

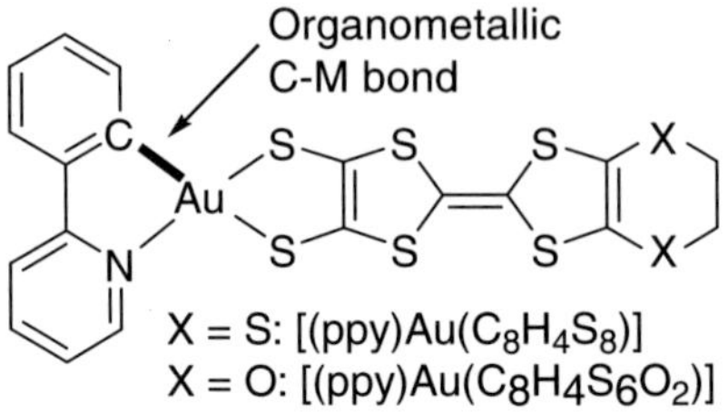

Scheme 1 Schematic drawing of unsymmetrical organometallic donors

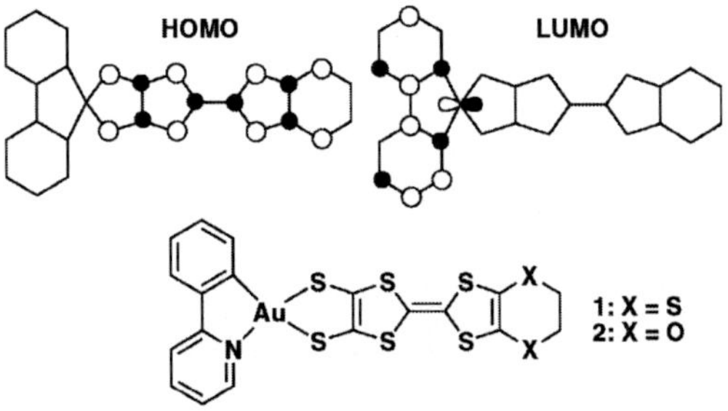

Fig. 4 HOMO and LUMO of the unsymmetrical organometallic gold-dithiolene complexes. (Reprinted with permission from [35]. Copyright 2008 American Chemical Society)

($C_8H_4S_8$ or $C_8H_4S_8O_6$)] are not similar to that of [(bpy)Pt($C_8H_4S_8$)] due to the orthometalation. In addition, the energy levels of the ground states for the orthometalated complexes are lower than those of the related diimine complexes [26, 27]. First redox potentials and HOMO-LUMO gaps of [(bpy)Pt($C_8H_4S_8$)] and [(ppy)Au($C_8H_4S_8$)] are summarized in Table 2. Donor ability of the organometallic gold complex is lower than that of the bpy-platinum complex. The HOMO-LUMO gap of the organometallic donor is larger than that of the bpy-platinum complex [14, 26]. Therefore, the σ-bond coordination markedly affects the HOMO and LUMO energy levels on the complexes. The organometallic Au complexes can form air stable crystal (Fig. 5), and provide various single crystals of their 2:1 cation radical salts by use of electrochemical crystallization [26, 27]. The 2:1 salts are more advantageous for improvement of their conducting properties than the 1:1 salts such as [(bpy)Pt($C_8H_4S_8$)][BF_4] (see Sect. 2.2.2) [22, 23].

3.2 *Cation Radical Salts of the Organometallic Au(III) Dithiolene Complexes*

Various cation radical salts [(ppy)Au(S-S)]$_2$[anion][solvent]$_n$ (S-S = $C_8H_4S_8$ or $C_8H_4S_6O_2$, anion = PF_6^-, BF_4^-, AsF_6^-, TaF_6^-, solvent = PhCl, n = 0–0.5) were prepared by constant current electrolysis of their benzonitrile or chlorobenzene solutions containing (Bu_4N)(anion) as electrolyte [26, 27, 35, 37].

Table 2 First redox potentials and HOMO-LUMO gaps of [(bpy)Pt($C_8H_4S_8$)] and [(ppy)Au($C_8H_4S_8$)]

	[(bpy)Pt($C_8H_4S_8$)]	[(ppy)Au($C_8H_4S_8$)]
First redox potential (V vs Ag/Ag^+)	−0.2	+0.09
HOMO-LUMO gap (Ev)	2.1	2.7

3.2.1 Crystal Structures

[(ppy)Au($C_8H_4S_8$)]$_4$[anion]$_2$[PhCl]: anion = AsF_6^-, TaF_6^-

The packing diagram of [(ppy)Au($C_8H_4S_8$)]$_4$[AsF_6]$_2$[PhCl] is shown in Fig. 6 This crystal contains four crystallographically independent cation radicals, two octahedral anions and a solvent molecule in the asymmetric unit [27]. The cation radicals form tetramers with a head-to-head configuration, and the crystal structure consists of stacking of the tetramers along the *a–c* direction with alternate orientations. There are some S···S nonbonded contacts (3.61–3.68 Å) between the cation moieties (Fig. 6b). Solvent molecules, PhCl and counter ions, AsF_6^- are situated between the cation columns (Fig. 6a). The crystal structure of the TaF_6 salt is similar to that of the AsF_6 salt [27].

[(ppy)Au($C_8H_4S_8$)]$_2$[PF_6] and [(ppy)Au($C_8H_4S_8O_6$)]$_2$[BF_4]

The cation radical salts [(ppy)Au($C_8H_4S_8$)]$_2$[PF_6] (PF_6 salt) and [(ppy)Au($C_8H_4S_8O_6$)]$_2$[BF_4] (BF_4 salt) exhibit different donor arrangements from those in the AsF_6 and TaF_6 salts. For each crystal, the asymmetric unit contains two cations and one anion. Columnar structures are formed by twofold head-to-head stacking of the cation radicals in the manner ····ABAB···· (Fig 7a, b) [27, 35, 37]. Figure 7c, d shows side-views of the columns. Significant difference is observed in the distances between the molecular planes. This difference corresponds to the different overlapping modes in the PF_6 and BF_4 salts, as shown in Fig. 7e, f. The intermolecular interactions between oxygen and hydrogen atoms in the $C_8H_4S_6O_2$ moiety are observed along the stacking direction in the BF_4 salt (Fig. 7d) [38]. For the PF_6 salt, many intermolecular S···S contacts (3.339–3.616 Å) shorter than the van der Waals distance (<3.7 Å) are observed within the column, as well as between the columns (Fig. 7a). However, only the intercolumn S···S contacts are observed in the BF_4 salt (3.686 Å, Fig. 7b).

3.2.2 Electrical Properties of the Cation Radical Salts

[(ppy)Au($C_8H_4S_8$)]$_4$[anion]$_2$[PhCl]: anion = AsF_6^-, TaF_6^-

The AsF_6 and TaF_6 salts exhibit semiconducting behavior ($\rho_{r.t.} = 4.0 \times 10^2$–$4.0 \times 10^3$ Ω cm; E_a = 0.11–0.15 eV) at ambient pressure. In the crystals, there are some

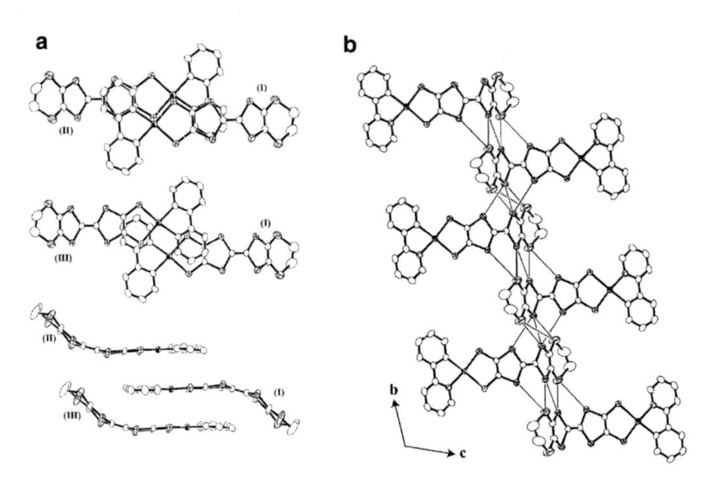

Fig. 5 (**a**) Molecular arrangements viewed along the molecular planes. (**b**) Packing diagram viewed along the *bc* plane in [(ppy)Au($C_8H_4S_8$)]. *Fine lines* show sulfur-sulfur contacts (<3.7 Å) (Reprinted with permission from [26]. Copyright 2003 Elsevier)

S···S contacts. However, conducting pathways are not formed effectively by the stacking of the cation radicals [26].

[(ppy)Au($C_8H_4S_8$)]$_2$[PF_6] and [(ppy)Au($C_8H_4S_8O_6$)]$_2$[BF_4]

The PF_6 salt exhibits semiconductive behavior with a small activation energy ($\rho_{r.t.} = 2.6\ \Omega$ cm; $E_a = 0.03$ eV) at ambient pressure. The small activation energy suggests that the PF_6 salt is situated close to the metallic state (see Sect. 3.2.3.2). The application of pressure is an effective method to enhance the intermolecular interactions and the band width. Indeed, metallic behavior appears above 0.8 GPa. In addition, at 1.6 GPa, the metallic region reaches down to 20 K (Fig. 8). To our knowledge, the PF_6 salt is the first example of the metallic organometallic compound, even though pressure is required to achieve this state [35]. On the other hand, the BF_4 salt is highly insulating ($\rho_{r.t.} > 1 \times 10^5\ \Omega$ cm) at ambient pressure and room temperature.

3.2.3 Electronic States of the Cation Radical Salts of Organometallic Metal–Dithiolene Complexes

Energy Band Structures of the Cation Radical Salts [(ppy)Au(S-S)]$_2$[anion] (S-S = $C_8H_4S_8$ or $C_8H_4S_6O_2$, anion =PF_6^- or BF_4^-)

Energy band calculation, spectroscopic and magnetic measurements are powerful methods to investigate the electronic states of the molecular conductors. Figure 9 shows energy band structures of the cation radical salts [(ppy)Au($C_8H_4S_8$)]$_2$[PF_6] (PF_6 salt) and [(ppy)Au($C_8H_4S_8O_6$)]$_2$[BF_4] (BF_4 salt) calculated by the tight-

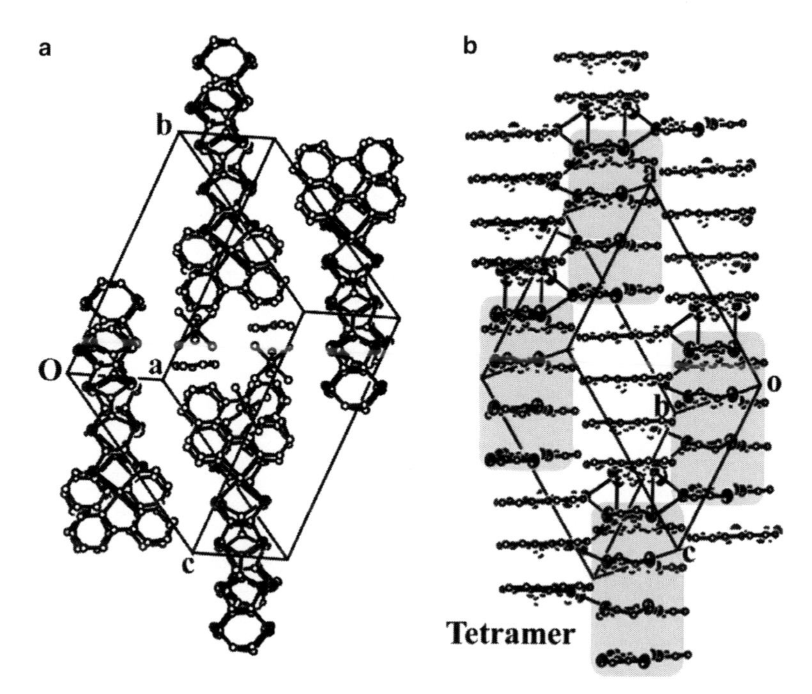

Fig. 6 (**a**) Crystal structure of $[(ppy)Au(C_8H_4S_8)]_4[AsF_6]_2[PhCl]$ along the molecular stacking direction. (**b**) End-on view of cation moieties of $[(ppy)Au(C_8H_4S_8)]_4[AsF_6]_2[PhCl]$. *Fine lines* indicate sulfur-sulfur contacts shorter than the van der Waals radii (<3.7 Å). (Reprinted with permission from [27]. Copyright 2005 Elsevier)

binding method with the extended Hückel MO calculation on the basis of the structural data [35]. The results of the calculations show that the difference in the cation arrangements provides remarkable distinctions of the electronic structure. The overlap integrals between HOMOs indicate strong dimerization in the BF_4 salt, while the cations are weakly dimerized in the PF_6 salt (Table 3). The band calculations suggest that the PF_6 salt has a one-dimensional metallic band structure with two pairs of planar Fermi surfaces. On the other hand, the BF_4 salt is a band insulator. This result is consistent with the resistivity measurements of the BF_4 salt. Thus, the BF_4 salt is a band insulator. However, the band calculation does not agree with the electrical resistivity measurement for the PF_6 salt which is due to the strong electron correlation in the salt as discussed in the next section [35].

Electronic State of $[(ppy)Au(C_8H_4S_8)]_2[PF_6]$

Temperature dependence of magnetic susceptibility of the PF_6 salt was measured from 300 to 4 K at 5 T [35]. The spin susceptibility of this salt gradually decreases from 300 to 50 K. Below 50 K, the susceptibility exhibits a rapid decrease accompanied by anisotropic temperature dependence, which is an indication of the long-range antiferromagnetic ordering. A one-dimensional Heisenberg model is

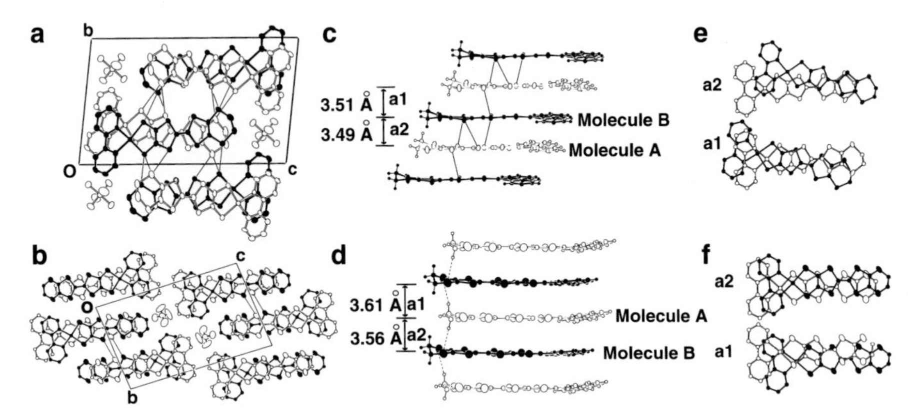

Fig. 7 Packing diagrams of (**a**) [(ppy)Au($C_8H_4S_8$)]$_2$[PF_6] (PF_6 salt) and (**b**) [(ppy)Au($C_8H_4S_6O_2$)]$_2$[BF_4] (BF_4 salt) viewed along the *a* axis. Side-views of the columns in the (**c**) PF_6 salt and (**d**) BF_4 salt. Molecular arrangements of two crystallographically independent molecules within the column of (**e**) PF_6 salt and (**f**) BF_4 salt. *Fine lines* indicate S···S contacts shorter than 3.7 Å. *Dashed lines* indicate O···H contacts within the range of 2.62–2.70 Å. (Reprinted with permission from [35]. Copyright 2008 American Chemical Society)

applicable to the data above 100 K. The reflectance spectrum indicates an evidently dimeric structure. From the Raman and IR measurements, no charge ordering was found [35]. All these results indicate that the PF_6 salt is an effectively half-filled system due to the (weak) dimerization and a paramagnetic insulator with a localized S = 1/2 spin on each dimer at ambient pressure (Fig. 10). That is, this system is a quasi-one-dimensional Mott insulator based on the dimer unit of uniformly charged donor molecules with a site-charge of +0.5. This means that the band crossing of the second and third subbands (ii) and (iii) in Fig. 9a is an artifact of the calculation and the actual band structure has a gap between subbands (ii) and (iii). However, this artifact suggests that the dimerization gap is expected to be narrow and the antiferromagnetic Mott insulating state is situated in the vicinity of the metal-insulator boundary [35]. One possible mechanism of the pressure-induced metallic state is a removal of the effective half-filled state. If the subbands (ii) and (iii) hybridize well with each other under pressure, the effectively half-filled state turns to the quarter-filled state with wider band width.

3.3 Conclusion

New molecular architecture of the components for molecular conductors was developed by using the unsymmetrical organometallic-dithiolene complexes which were modified from the unsymmetrical diimine-dithiolene complexes by the introduction of carbon-metal σ-bond. Two organometallic donor components with $C_8H_4S_8$ and $C_8H_4S_6O_2$ as the dithiolene ligands were synthesized and their 2:1 cation radical salts [(ppy)Au(S-S)]$_2$[anion][solvent]$_n$ (S-S =$C_8H_4S_8$ or $C_8H_4S_6O_2$, anion = PF_6^-, BF_4^-, AsF_6^-, TaF_6^-, solvent = PhCl, n = 0–0.5) were prepared as

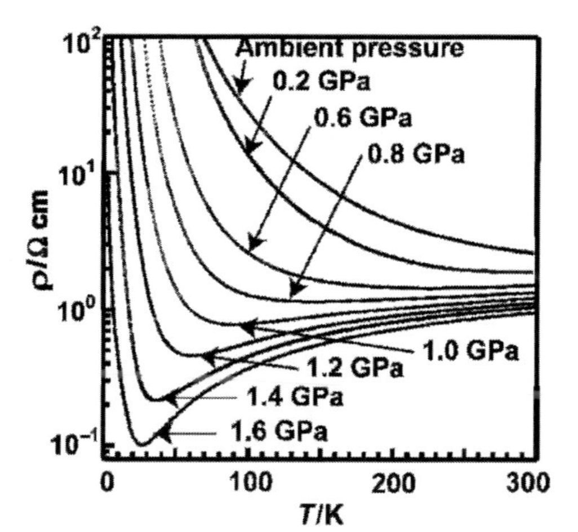

Fig. 8 Temperature dependence of the resistivity for $[(ppy)Au(C_8H_4S_8)]_2[PF_6]$ at various pressures form ambient pressure to 1.6 GPa. (Reprinted with permission from [35]. Copyright 2008 American Chemical Society)

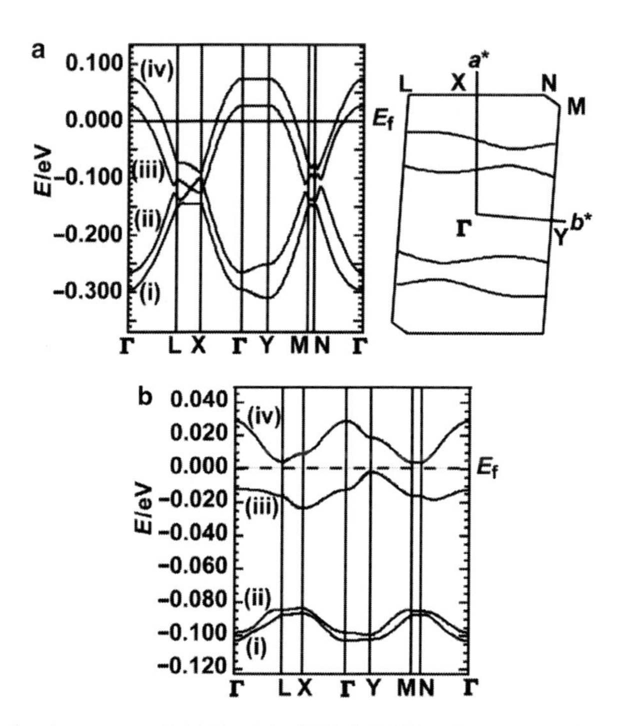

Fig. 9 Energy band structures of: (**a**) $[(ppy)Au(C_8H_4S_8)]_2[PF_6]$; (**b**) $[(ppy)Au(C_8H_4S_8O_6)]_2[BF_4]$. (Reprinted with permission from [35]. Copyright 2008 American Chemical Society)

single crystals by electrochemical crystallization. $[(ppy)Au(C_8H_4S_8)]_2[PF_6]$ under pressure is the first molecular metal based on the organometallic component. At ambient pressure, this system is the Mott insulator in the vicinity of the metal-insulator boundary. Other salts are band insulators.

Table 3 Calculated overlap integrals (S) between HOMOs for the cation radical salts the PF_6 and BF_4 salts. (Reprinted with permission from [35]. Copyright 2008 American Chemical Society)

S ($\times 10^{-3}$)	PF_6 salt	BF_4 salts
a1	−8.3	−0.8
a2	−9.0	−4.6
b1	1.3	−0.2
b2	−1.9	−0.4
b3	1.1	−1.5
b4	−1.2	−0.2
p1	−1.2	−0.7
p2	−0.9	−0.2

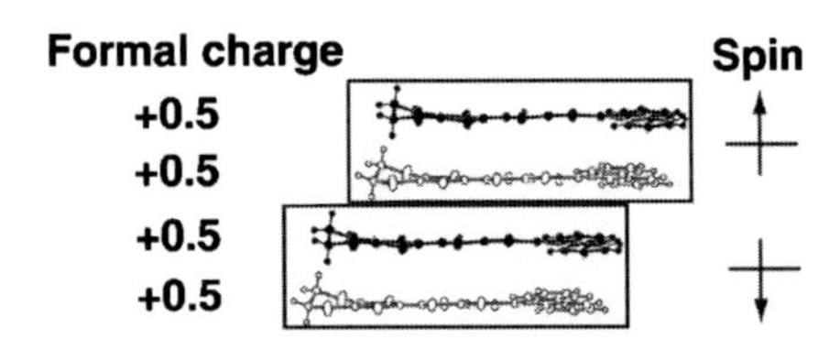

Fig. 10 Schematic display of the electronic structure of $[(ppy)Au(C_8H_4S_8)]_2[PF_6]$ at ambient pressure (Reprinted with permission from [35]. Copyright 2008 American Chemical Society)

4 Other Unsymmetrical Organometallic–Dithiolene Complexes for the Components of Molecular Conductors

The molecular charge of the organometallic complexes shown in Scheme 1 can be modified easily by the choice of the metal ions. Suga et al. prepared monovalent organometallic platinum-dithiolene complexes which have the same molecular structure with that of the gold complexes (Figs. 11 and 12), and succeeded in obtaining neutral radicals of the organometallic platinum-dithiolene complex by chemical oxidation [39]. These neutral radicals can be expected to develop toward new type of single component molecular conductors [3], although these neutral radicals are insulators and their crystal structures have not been revealed yet.

5 General Conclusion

Chemical research of molecular metals was activated by the discovery of the metallic charge transfer salt TTF-TCNQ in 1973 [40]. Two basic molecular architectures have been studied intensively. One is based on organic molecules with the

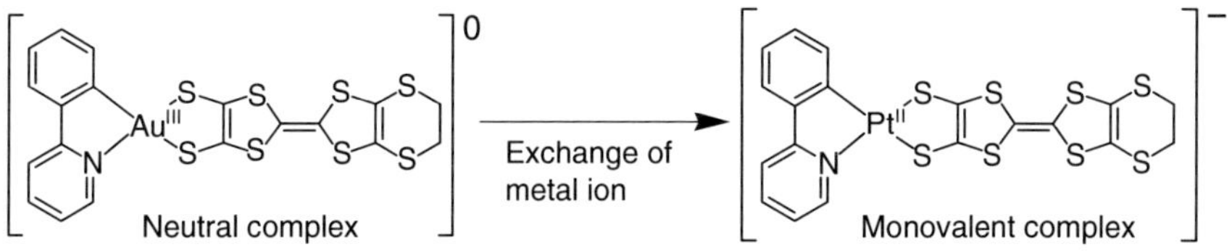

Fig. 11 Schematic drawing of neutral and monovalent complexes based on the organometallic-dithiolene compounds

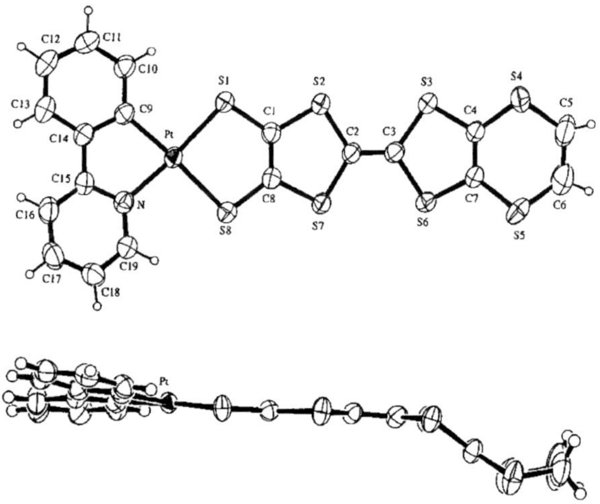

Fig. 12 Anion moiety in $(Bu_4N)[(ppy)Pt(C_8H_4S_8)]$. (Reprinted with permission from [39]. Copyright 2004 Chemical Society of Japan)

TTF moiety such as BEDT-TTF. The other is based on metal-dithiolene complexes such as $M(dmit)_2$. Both architectures succeeded in providing various highly conducting materials which exhibit superconducting and metallic behavior. Researches of the molecular conductors have focused on the fundamental aspects of the science of molecular conductors. Recently, however, the molecular conductors have been applied to functional materials such as nano devises. New molecular architectures for control of crystal structures and electronic properties have been required. Matsubayashi et al. proposed new molecular architecture by using the unsymmetrical diimine-dithiolene complexes. They have established the basic chemistry of the complexes by spectroscopic and electrochemical measurements and revealed conducting properties of the salt containing the unsymmetrical complexes, although the single crystals were not obtained. Modification of the unsymmetrical molecules by the introduction of the carbon-metal σ-bond was examined in order to improve crystal quality and the conducting ability of their cation radical salts. Kubo et al. have succeeded in preparation of the first organometallic components by the modification of the unsymmetrical complexes and their 2:1cation radical salts as single crystals. Among them, the cation radical salt $[(ppy)Au(C_8H_4S_8)]_2[PF_6]$ is the first metallic compound derived from organometallic compounds, even though pressure is required to achieve the metallic state. This has opened the way to the basic science of the electrically conducting organometallic compounds. This type of organometallic complex can easily change their molecular charge by the choice of the ligands and metal ions. Suga et al. have succeeded in preparation of neutral radicals based on the unsymmetrical platinum dithiolene complexes for new type of single components molecular conductors. Although the number of reports on molecular conductors derived from the organometallic complexes is limited at present, the organometallic compounds would be the vast frontiers of molecular conductors.

References

1. Akamatu H, Inokuchi H, Matsunaga Y (1954) Nature 173:168
2. Mori T, Kobayashi A, Sasaki Y, Kobayashi H, Saito G, Inokuchi H (1984) Bull Chem Soc Jpn 57:627
3. Tanaka H, Okano Y, Kobayashi H, Suzuki W, Kobayashi A (2001) Science 291:285
4. Kagoshima S, Kato R, Fukuyama H, Seo H, Kino H (1999) In: Bernier P, Lefrant S, Bidan G (eds) Advances in synthetic metals – twenty years of progress in science and technology. Elsevier Science, Amsterdam, p 262
5. Kato R (2000) Bull Chem Soc Jpn 73:515
6. Miller JS, Epstein AJ (1976) Prog Inorg Chem 20:1
7. Kato R (2004) Chem Rev 104:5319 and references cited therein
8. Underhill AE, Ahmad MM (1981) J Chem Soc Chem Commun 67
9. Canadell E, Ravy S, Pouget JP, Brossard L (1990) Solid State Commun 75:633
10. Watanabe E, Fujiwara M, Yamaura JI, Kato R (2001) J Mater Chem 11:2131
11. Matsubayashi G, Yamaguchi Y, Tanaka T (1988) J Chem Soc Dalton Trans 2215
12. Matsubayashi G, Hirao M, Tanaka T (1988) Inorg Chim Acta 144:217
13. Nakahama A, Nakano M, Matsubayashi G (1999) Inorg Chim Acta 284:55
14. Kubo K, Nakano M, Tamura H, Matsubayashi G (2000) Inorg Chim Acta 311:6
15. Miller TR, Dance IG (1973) J Am Chem Soc 95:6970
16. Vogler A, Kunkely H, Hlavatsch J, Merz A (1984) Inorg Chem 23:506
17. Zuleta JA, Bevilacqua JM, Proseripio DM, Harvey PD, Eisenberg R (1992) Inorg Chem 31:2396
18. Paw W, Cummings SD, Mansour MA, Connick WB, Geiger DK, Eisenberg R (1998) Coord Chem Rev 171:125
19. Cocker TM, Bachman RE (2001) Inorg Chem 40:1550
20. Makedonas C, Mitsopoulou CA, Lahoz FJ, Balana AI (2003) Inorg Chem 42:8853
21. Chen CT, Liao SY, Lin KJ, Chen CH, Lin TYJ (1999) Inorg Chem 38:2734
22. Shibaeva RP, Yagubskii EB (2004) Chem Rev 104:5347 and references cited therein
23. Kobayashi H, Cui HB, Kobayashi A (2004) Chem Rev 104:5265 and references cited therein
24. Kubo K, Nakano M, Tamura H, Matsubayashi G (2002) Inorg Chim Acta 336:120
25. Matsubayashi G, Nakano M, Tamura H (2002) Coord Chem Rev 226:143
26. Kubo K, Nakano M, Tamura H, Matsubayashi G, Nakamoto M (2003) J Organomet Chem 669:141
27. Kubo K, Nakao A, Ishii Y, Kato R, Matsubayashi G (2005) Synth Met 153:425
28. Mdleleni MM, Bridgewater JS, Watts RJ, Ford PC (1995) Inorg Chem 34:2334

29. Schmid B, Garces FO, Watts RJ (1994) Inorg Chem 33:9
30. Craig CA, Watts RJ (1989) Inorg Chem 28:309
31. Ichimura K, Kobayashi T, King KA, Watts RJ (1987) J Phys Chem 91:6104
32. Ohsawa Y, Sprouse S, King KA, DeArmond MK, Hanck KW, Watts RJ (1987) J Phys Chem 91:1047
33. Sprouse S, King KA, Spellane PJ, Watts RJ (1984) J Am Chem Soc 106:6647
34. King KA, Spellane PJ, Watts RJ (1985) J Am Chem Soc 107:1431
35. Kubo K, Nakao A, Ishii Y, Yamamoto T, Tamura M, Kato R, Yakushi K, Matsubayashi M (2008) Inorg Chem 47:5495
36. Mansour MA, Lachicotte RJ, Gysling HJ, Eisenberg R (1998) Inorg Chem 37:4625
37. Kubo K, Nakano M, Tamura H, Matsubayashi G (2003) Eur J Inorg Chem 4093
38. Horiuchi S, Yamochi H, Saito G, Sakaguchi KI, Kusunoki M (1996) J Am Chem Soc 118:8604
39. Suga Y, Nakano M, Tamura H, Matsubayashi G (2004) Bull Chem Soc Jpn 77:1877
40. Ferraris J, Cowan DO, Walatka V, Perlstein JH (1973) J Am Chem Soc 95:948

Top Organomet Chem (2009) 27: 55–75

Electroactive Paramagnetic Complexes as Molecular Bricks for π–*d* Conducting Magnets

Stéphane Golhen, Olivier Cador, and Lahcène Ouahab

Abstract On one hand, tetrathiafulvalene and its derivatives are widely employed to promote electronic conductivity on the macroscopic scale because of their redox activity. On the other hand, metallic complexes are great sources of sophisticated magnetic materials. The meeting of these "two hands," when they interact, in other words when localized electrons (*d* spins) interact with itinerant electrons (π carriers), may give rise to a new type of multifunctional materials: magnetic and conducting. In the following, we develop the strategy we are applying to prepare π–*d* systems: that is the preparation of redox active complexes bearing redox active ligands, covalently linked to metal complexes. The synthesis and physical properties of mononuclear and polynuclear redox active paramagnetic complexes are presented. Particular attention is given to the packing of TTF moieties.

Keywords π-*d* interaction, *3d* transition metal, Electronic conductivity, Magnetism, Redox-active complexes, TTF

Contents

S.Golhen, O.Cador, and L.Ouahab(✉)
Organometallic and Molecular Materials, UMR 6226 CNRS-UR1 Sciences Chimiques de Rennes, Université de Rennes 1, 263 Avenue du Général Leclerc, 35042 Rennes Cedex, France, E-mail: lahcene.ouahab@univ-rennes1.fr

M. Fourmigué and L. Ouahab (eds.), *Conducting and Magnetic Organometallic Molecular Materials*, Topics in Organometallic Chemistry 27,
DOI: 10.1007/978-3-642-00408-7_3,

1 Introduction

Both words magnetism and electricity have fascinated mankind for a thousand years. From the etymological sense the origin of electricity is well established; it comes from the yellow amber (electron in Greek). In contrast, the origin of the word magnetism is not so clear and several stories have been proposed. Amongst them, the most interesting is probably the story of shepherd *Magnes* who noticed that some stones (the magnetite Fe_3O_4) stuck to sandal nails. Nowadays, magnetism is still surrounded by mystery and may be employed, for example, to describe the faculty for some human beings to focus the attention. For the scientist, of course, magnetism is not so obscure and comes from electron spin and/or movement of electrons.

Electronic and magnetic materials have developed widely in the second half of the twentieth century and are now omnipresent in daily life devices. In the last two decades the giant magnetoresistance phenomenon showed how electrical conductivity and magnetism can be intimately correlated inside matter. Most of the commercially available devices are built on the basis of solid-state chemistry. In parallel with the development of new electronic and magnetic tools, molecular and supramolecular chemistry have started to produce semiconductors [1], conductors [2, 3], superconductors [4–6] and magnets [7, 8], employing molecular bricks as construction units. More recently, magnetic properties (coercivity and remanence) have been discovered on the molecular scale [9], the so-called single-molecule magnets (SMMs) and, in one-dimensional assemblies [10, 11], the so-called single-chain magnets (SCMs). However, in all these new materials the electronic properties (conduction or magnetism) are considered to be separate. In other words, these materials possess a single function which follows one single electronic property.

In the last decade the scientific community specialized in the design and studies of molecular-based materials has searched to combine, at the molecular level, different physical properties in a single material to produce multifunctional materials. Some important results are the coexistence of paramagnetism and conductivity [12], paramagnetism and superconductivity [13–15], antiferromagnetism and conductivity [16], and perhaps one of the most striking results has been the preparation of conducting magnets [17]. These hybrid materials result from the stacking of two parallel networks (organic and inorganic). The organic network is made of electroactive molecules derived from tetrathiafulvalene (TTF), namely bis(ethylenedithio) tetrathiafulvalene (BEDT-TTF) (Fig. 1).

Such molecules possess the specific property to stack in partially oxidized form in the solid state, and build networks into which the π electrons of the donors are delocalized and promote the metallic state. The inorganic network is a 2D bimetallic complex into which the localized spins of transition metal ions are coupled via superexchange pathway through oxalate bridges. The particularity of this 2D network is that it orders ferromagnetically (magnet) at low temperature. In the low temperature range (i.e., below 5.5 K) the material behaves as a metallic magnet. However, from the electronic point of view the two lattices behave quasi-independently. While the

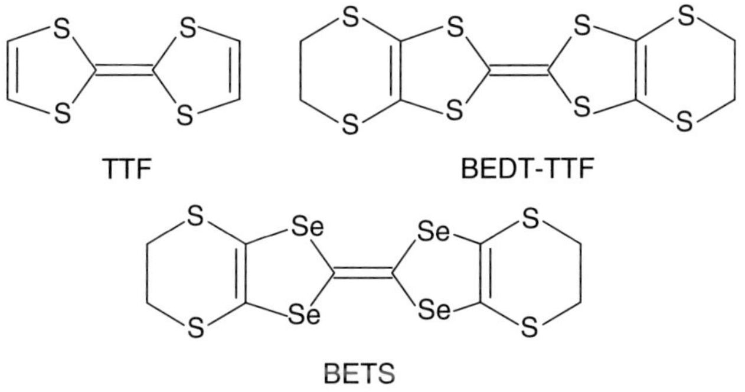

Fig. 1 Chemical structures of TTF, BEDT-TTF and BETS electroactive donor molecules

conducting electrons have no influence on the magnetic properties, the influence of the inorganic network on the electrical conductivity occurs only through the internal field created by the ferromagnetic layers. There is no communication at the molecular level between the conducting and the magnetic electrons.

One of the current challenges is to establish communication between the networks through so-called π–*d* interactions. Numerous results have been obtained worldwide [18, 19] with the demonstration of the coexistence of antiferromagnetic and superconducting state [20, 21], and field-induced superconductivity [22, 23]. Indeed, the antiferromagnetic ordering of the magnetic moments of the inorganic network is the consequence of the three-dimensionality of the magnetic connections. More strictly speaking, the interactions (exchange or superexchange) between the spins propagate in the three space directions. The common point of all these materials is that the π-*d* interaction takes place through space. Indeed, crystal structures are constituted of alternate stacking of an inorganic network made of $[Fe^{III}X_4]^-$ tetrahedra (X = Br [21], Cl [22]) and of an organic network in which the partially oxidized donors BETS (bis(ethylenedithio)tetraselenafulvalene) are dimerized and organized in κ-type [21] or λ-type [22] arrangement. Close contacts between atoms belonging to the two networks switch on π–*d* interactions.

Basically, two conditions must be fulfilled to establish through space π–*d* interactions:

1. First, close contacts, typically of the range of the sum of the van der Waals radii of the atoms involved, or hydrogen bonds, must exist between the organic network and the inorganic moieties. It must be mentioned that interactions between the localized spins in a hypothetical inorganic network are not necessary at this stage; if one is seeking magnetic three-dimensional ordering the only requirement is that π–*d* interactions propagate in 3D. In fact, as illustrated on Fig. 2, the π–*d* interactions polarize the localized spins. This is the supramolecular equivalent of the RKKY (Ruderman-Kittel-Kasuya-Yosida) coupling [24] where instead of an intraatomic coupling between localized (*d* or *f*) spins and *s* conducting electrons the coupling takes place between chemically independent molecules.

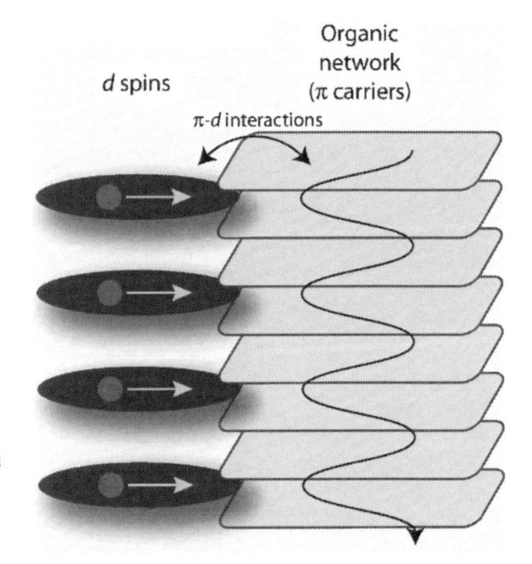

Fig. 2 Representation of the ferromagnetic alignment of *d* localized spins resulting from the interaction with itinerant electrons through short intermolecular contacts

2. Second, the spin density maps must cover the atoms which are involved in the contacts. On the one hand, the localized *d* spins of the transition metal ions must paradoxically be sufficiently delocalized toward the ligand atoms. On the other hand, the π carriers of the donor set have to extend toward the peripheral atoms. In fact the electrons belonging to the two networks must meet at the intermolecular contact, otherwise the networks would ignore each other.

The coexistence of these two conditions is a subtle balance between supramolecular structuration and spin delocalization. As an example, iodine substituted donors [25–29] can easily interact with nitrogen-containing [19, 30] or sulfur-containing [31] ligands transition metal complexes. Typical contacts between iodine atoms of TTFs derivatives (DIET = diiodo-(ethylenedithio)tetrathiafulvalene) and sulfur atoms of isothiocyanoto groups of complexes $[Cr^{III}(isoq)_2(NCS)_4]^-$ (isoq = isoquinoline) range from 3.25 to 3.5 Å. These contacts are remarkably shorter (Fig. 3) than the sum of the van der Waals radii of the two atoms (1.8 Å(S) + 1.98 Å(I) = 3.78 Å). The hybrid materials are then well-structured (Fig. 3) with a one-dimensional array of the partially oxidized (charge 1/2+) donors[1] and a one-dimensional array of the inorganic anions $[Cr^{III}(isoq)_2(NCS)_4]^-$ thanks to π–π overlaps between adjacent isoq rings. Spin density calculations performed with the aid of DFT method using the Amsterdam Density Functional (ADF) program [32] (developed by Baerends and coworkers [33, 34]) showed that 3/2-spin mainly centered on Cr^{III} partially delocalizes toward *S*-atoms of the four isothiocyanoto groups. The spin density on sulfur atoms being equal to ~0.1 while it is equal to 2.8 on Cr^{III}. Nevertheless, nearly no spin density (<0.004) is found on terminal iodine atoms of $(DIET)_2$. Then only very weak interaction is expected between localized and delocalized spins and

[1]The donors are in fact organized in + 1 charged dimers which stack to form regular chain

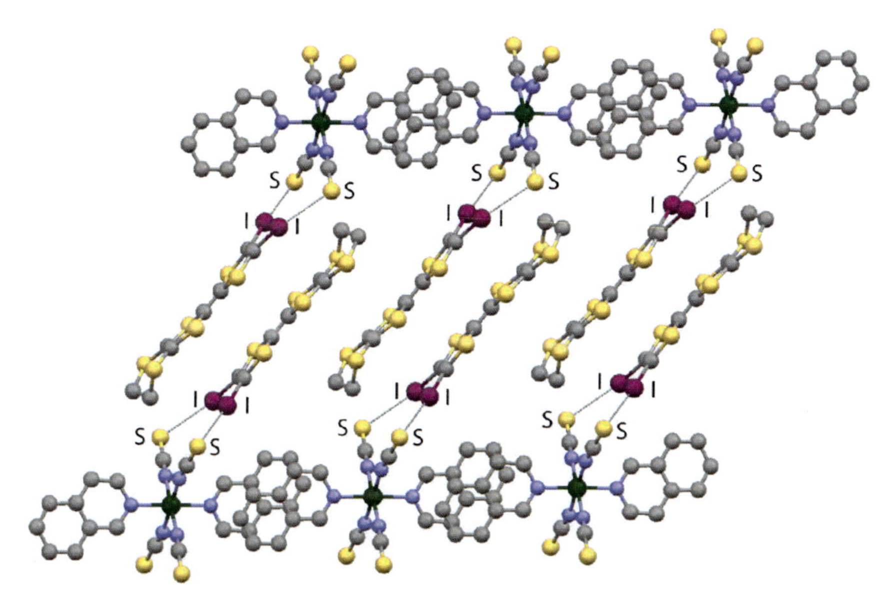

Fig. 3 Structural arrangement in $(DIET)_2[Cr^{III}(isoq)_2(NCS)_4]$ (phase *a*) of the inorganic and organic moieties showing the one-dimensional arrays and the remarkably short intermolecular contacts (thin gray line) between iodine-substituted donors and sulfur atoms of the isothiocyanato ligands

that is indeed the case. In fact, the formation of compound can be viewed as a double autoassembling of two ingredients, the organic donor and the paramagnetic complexes, to form (1) the conducting layers and (2) the hybrid materials with through-space π–*d* interactions.

One must say that perhaps the major challenge in this so-called *through space* strategy is to enhance the strength of the π–*d* interactions. Until now and to the best of our knowledge the amplitude of such interactions in TTF-based compounds remains very small. It is of the order of the wavenumber at the exception of few examples, which are, and this is the drawback, insulators [35–38]. An alternative has emerged in the last few years. The philosophy is to design paramagnetic complexes functionalized by electroactive ligands derived from TTF.

These electroactive paramagnetic units are then the building blocks of edifices in which a synergy between magnetism and electrical conductivity is established. Several arguments plead in favor of this so-called *through bond* strategy (Fig. 4):

1. The donor is covalently linked to the paramagnetic moiety and then stronger interactions are expected between the organic centered radicals and the paramagnetic centers. Of course, this statement is true only if the functionalized donors are fully conjugated, or, in other words, if the TTF radicals spread over the chelating site(s).

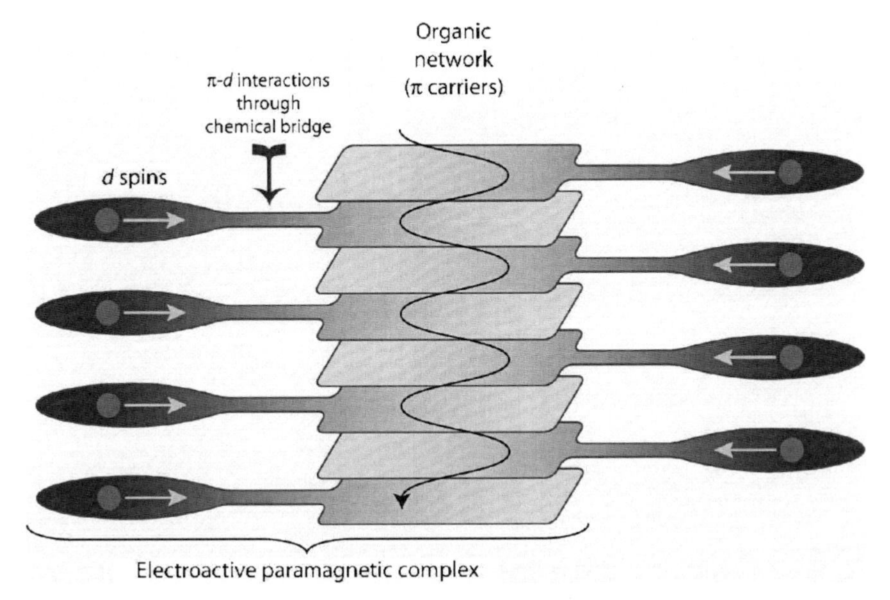

Fig. 4 Representation of the supramolecular arrangement of electroactive paramagnetic complexes emphasizing the π–*d* interactions through a chemical bridge between electronic spins of TTFs (or BEDT-TTFs) and *d* spins of transition metal ions and the polarization effects on the magnetic moment alignments of each complex

2. The supramolecular game involves only one brick instead of two, so the number of degrees of freedom is reduced. The autoassembling process which takes place between π-carriers governs the final three-dimensional architecture of the material.
3. The complex itself may contain all the ingredients to act as an electroactive SMM if the inorganic moiety is selected in the SMMs library [39]. The complex is an isolated unit which might be directly employed as a molecular component in spintronic and/or electronic devices.

The primary molecules (TTF or BEDT-TTF) have, as a first step, to be chemically modified to enter the coordination sphere of paramagnetic transition metal complexes. The TTF chemistry is so rich that this book would not be enough to cover the different substitutions performed by organic chemists [40]. The second step is to fix those modified electroactive ligands onto mononuclear or polynuclear cores. We arbitrarily restrain ourselves to those which have been successfully coordinated to paramagnetic transition metal complexes. A great deal of work has been accomplished by numerous research groups to coordinate substituted TTFs or BEDT-TTFs by pyridine-type heterocycles [41–54], acetylacetonate ligands [55–57] as well as salen-type ligands [58].

In this chapter, we will focus on paramagnetic materials based on two pyridine substituted TTFs (Fig. 5): the TTF−CH=CH−py [59] and its trimethyl derivative Me_3TTF−CH=CH−py [51]. Owing to the presence of one pyridine group on each TTF, such ligands coordinate to one transition metal ion. Therefore, a large variety of transition metal complexes can, in principle, adapt pyridine substituted TTFs in

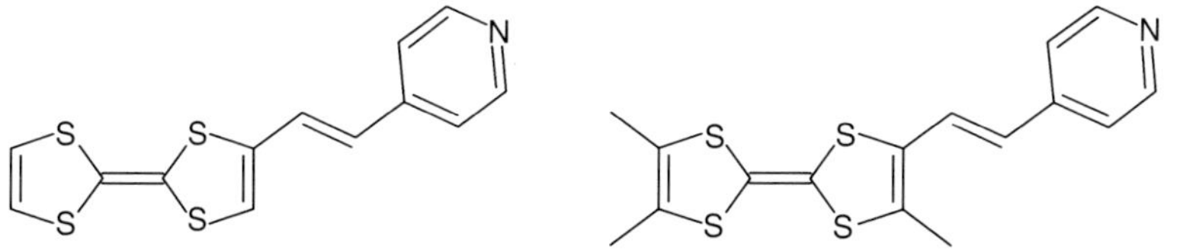

TTF-CH=CH-py Me_3TTF-CH=CH-py

Fig. 5 Chemical structures of the two electroactive donor molecules functionalized with pyridine heterocycle

their coordination spheres. Depending on the accessibility of the coordination sites, their number and also the number of transition metal ions in the complexes, various electroactive paramagnetic complexes have been synthesized and characterized. The *pseudo* stoichiometry (M:L) between the number of transition metal ions (M) and the number of coordinated electroactive ligands (L) varies from 1:1 to 3:2 with homometallic as well as heterometallic polynuclear complexes. The charge of the paramagnetic electroactive complexes reflects the formal charge of the electroactive ligands.

2 Mononuclear Compounds

2.1 Neutral Complexes

TTF–CH=CH–py was synthesized as described in the literature [59]. First compounds were obtained as mononuclear complexes by adding a hot *n*-hexane solution of TTF–CH=CH–py to a stirred boiling *n*-hexane solution of $M^{II}(hfac)_2\cdot H_2O$, with M = Mn, Cu. After stirring for 10 min, depending on the metal/ligand ratio (equal to 2 or 1), the resulting black crystals were characterized as $Cu^{II}(hfac)_2$(TTF–CH=CH–py) [45, 60] and *trans*- or *cis*- complex $M^{II}(hfac)_2$(TTF–CH=CH–py)$_2$ [49], for M = Cu, Mn respectively (see Fig. 6).

2.1.1 [$Cu^{II}(hfac)_2$(TTF–CH=CH–py)]

The coordination sphere of Cu^{II} is constituted of four oxygen atoms from two hfac ions and one nitrogen atom of the pyridine group of one TTF–CH=CH–py, making a pentacoordinated Cu^{II}. The crystal structure of this compound is reminiscent of what is usually observed for conducting radical-ion salts containing organic donors and inorganic anions. TTFs form an organic sublattice ((001) planes) whereas inorganic layers of [$Cu^{II}(hfac)_2$] are lying in the (002) planes. These two layers alternate and are bridged by conjugated CH=CH–py (see Fig. 7). The TTF moieties in the organic layers are parallel to each other. However no S···S contacts shorter than the sum of the van der Waals (vdw) radii of sulfur atoms are observed.

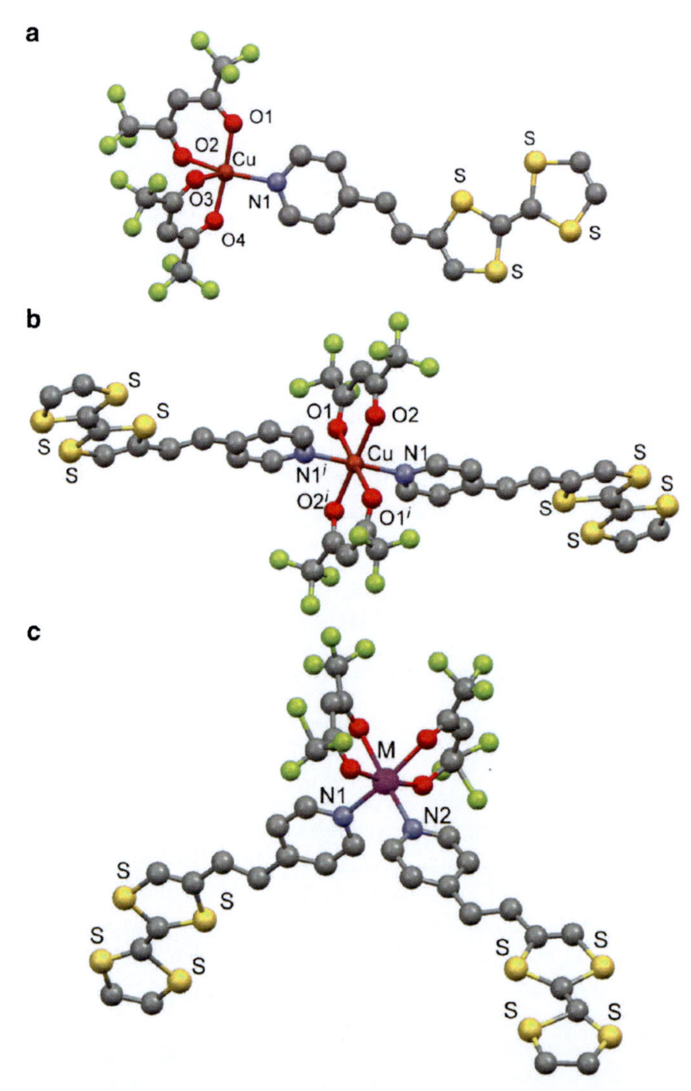

Fig. 6 Three conformations of neutral complexes. **(a)** $Cu^{II}(hfac)_2$(TTF−CH=CH−py)[2]. **(b)** *trans*-$Cu^{II}(hfac)_2$(TTF−CH=CH−py)$_2$ [48]. **(c)** *cis*-$M^{II}(hfac)_2$(TTF−CH=CH−py)$_2$ [49] with M = Cu, Mn. Hydrogen atoms are omitted for clarity

2.1.2 $M^{II}(hfac)_2$(TTF−CH=CH−py)$_2$, M = Cu and Mn

Neutral *trans*- and *cis*- complexes were obtained for Cu^{II} and Mn^{II} ions (Fig. 6b,c) respectively [49]. The metallic ion in octahedral environment is coordinated to four oxygen atoms of two hfac ions and two nitrogen atoms of two TTF−CH=CH−py

[2]CIF file is available on request from the authors

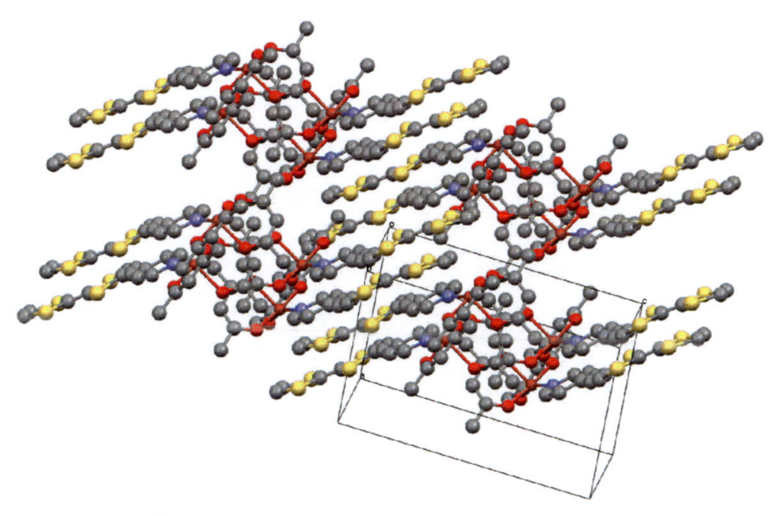

Fig. 7 Packing of $Cu^{II}(hfac)_2$(TTF−CH=CH−py) complexes showing the alternating inorganic and organic layers

ligands. The bond lengths and bond angles of the TTF moiety are close to those reported for the noncoordinated neutral unit, indicating, additionally to the charge of $M^{II}(hfac)_2$ that the TTF−CH=CH−py ligands are neutral. Like in the previous compound, the crystal structures consist of alternating organic (TTF−CH=CH−py) and inorganic $M^{II}(hfac)_2$ layers.

Cis- coordination

The chevron-patterned complexes Mn1 and Mn2 alternate along the [120] direction and form into the (110) plane, packed gutter-like motif. This "V" motif alternates with a similar reversed "Λ" motif along the *c* direction (Fig. 8). The crystal structure consists of TTF−CH=CH−py organic layers and inorganic layers of $Mn^{II}(hfac)_2$ alternating along the *c* axis. Three intermolecular S···S contacts shorter than the vdw radii are observed; they take place between two complexes shaped in the packing as "Λ" for one and "V" for the second related by inversion center.

Trans- Coordination

Trans- complex is obtained only with Cu^{II} which is coordinated to four oxygen atoms of two hfac ions and two nitrogen atoms of two TTF−CH=CH−py ligands. Cu^{II} lies on inversion center and therefore the TTF−CH=CH−py ligands are in *trans-* conformation. The copper ion adopts a Jahn-Teller distorted octahedral

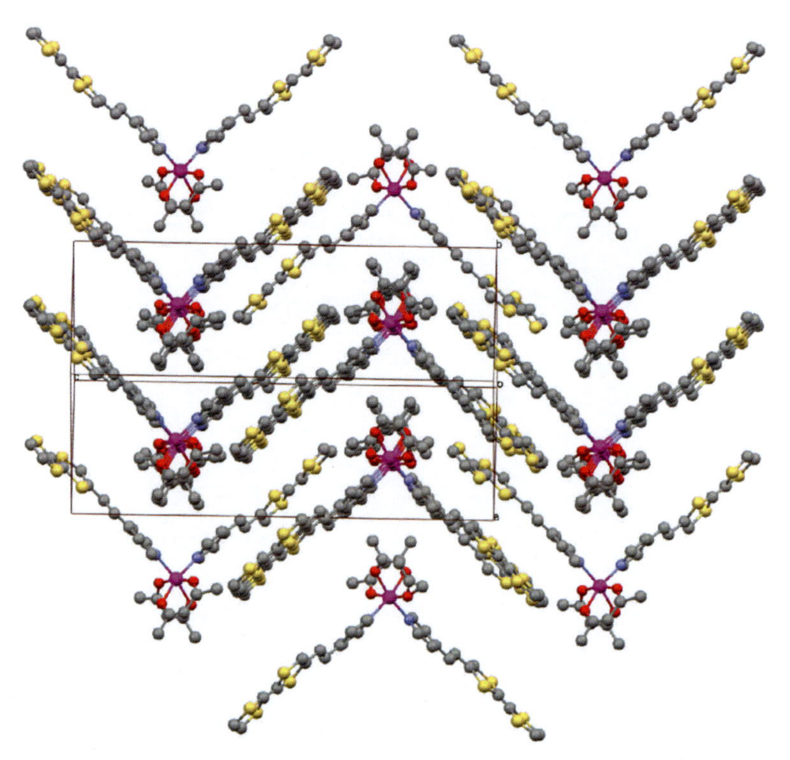

Fig. 8 Projection in the (110) plane of *cis*-$Mn^{II}(hfac)_2(TTF-CH{=}CH-py)_2$ showing alternating layers of organic and inorganic entities linked by conjugated bridge

coordination geometry. As previously, the crystal structure consists of alternating layers of organic TTF–CH=CH–py and inorganic $Cu^{II}(hfac)_2$ along *a* axis. The TTF–CH=CH–py molecules in the organic layers form orthogonal dimers. As expected, from the neutral form of electroactive ligands, the material is an insulator. Investigations of thin layer cyclic voltammetry (TLCV) reveal that two independent TTF–CH=CH–py ligands are simultaneously oxidized in one-electron process.

2.2 Radical Cation Salts $[M^{II}(hfac)_2(TTF-CH{=}CH-py)_2](A)_n{\cdot}Solv$ with $A = PF_6$, BF_4, $M = Cu$, Mn and $n = 1, 2$

From the neutral 1:2 complexes introduced before, two kinds of materials were obtained by electrocrystallization: fully oxidized salts where both TTF–CH=CH–py moieties are oxidized, and mixed valence salts in which the two ligands are partially oxidized. All oxidized complexes were obtained by oxidation at constant current of a neutral 1:2 complex in an electrolyte containing the anions, and using electrocrystallization technique with platinum wires (Ø = 1 mm) electrodes.

All crystal structures consist of alternating organic and inorganic layers linked by conjugated bridge. Focusing on the organic fragment of oxidized salts, one can distinguish two kinds of compounds: those where TTF moieties and anions are stacked together to built mixed layers, such as the following mixed valence complex *trans*-$[Cu^{II}(hfac)_2(TTF-CH{=}CH-py)_2](PF_6)$ [45] and those where organic layers are exclusively made of TTF moieties such as *trans*-$[Cu^{II}(hfac)_2(TTF-CH{=}CH-py)_2](BF_4)_2{\cdot}2CH_2Cl_2$ [61]. These two compounds are described below.

2.2.1 Fully Oxidized Complexes with TTF Layers

The complex *trans*-$[Cu^{II}(hfac)_2(TTF-CH{=}CH-py)_2](BF_4)_2{\cdot}2CH_2Cl_2$ was obtained after 1 week of galvanostatic oxidation of $Cu^{II}(hfac)_2(TTF-CH{=}CH-py)_2$ [61]. The molecular structure of the copper complex is identical to its neutral form. There is one TTF–CH=CH–py molecule per BF_4^- and one dichloromethane solvent molecule. The copper is located at the center of a centrosymetric-distorted octahedron; two TTF–CH=CH–py ligands in *trans*- conformation are bonded to Cu^{II} by the nitrogen atoms of the pyridyl rings. From the stoichiometry, the charge distribution corresponds to fully oxidized $TTF-CH{=}CH-py^{+\bullet}$ radical units.

The crystal structure consists of alternating layers of organic $TTF-CH{=}CH-py^{+\bullet}$ and inorganic $Cu(hfac)_2$ fragment linked by a conjugated bridge. TTF moieties are stacked along the [100] direction: two TTF fragments from two different $[Cu(hfac)_2(TTF-CH{=}CH-py)_2]^{2+}$ molecules form a dimer with a π-π stacking (shortest S···S contact equal to 3.289(2) Å) and two neighboring dimers are slightly shifted so there is only one short S···S contact (equal to 3.674(2) Å) between dimers (see Fig. 9a). Despite no short contacts being observed between the TTF columns, this packing can be described as planes parallel to (002) direction (see Fig. 9b). The most important feature to notice in this packing is that neither solvent molecules nor BF_4^- anions are localized in the organic chain.

2.2.2 Mixed Valence Compounds with Mixed TTF/Anion Layers

The asymmetric unit contains one $[Cu^{II}(hfac)_2(TTF-CH{=}CH-py)_2]^+$ radical cation and one PF_6^- anion both in special positions, as well as one dichloromethane solvent molecule in the general position. The Cu^{II} ion lies in a distorted octahedral coordination, two oxygen from two hfac ions and two nitrogen from TTF–CH=CH–py are *trans* to each other and form the square plane (Cu-N bond length of 2.016(5) Å); the two reminiscent oxygen atoms from the two hfac ions occupy the apical positions.

From the stoichiometry, the charge distribution gives rise to the complex $[Cu^{II}(hfac)_2(TTF-CH{=}CH-py)_2]^+$ complex; also, formally +1 charged $(TTF-CH{=}CH-py)_2^+$ dimers are found in the crystal lattice. *trans*-$[Cu^{II}(hfac)_2$

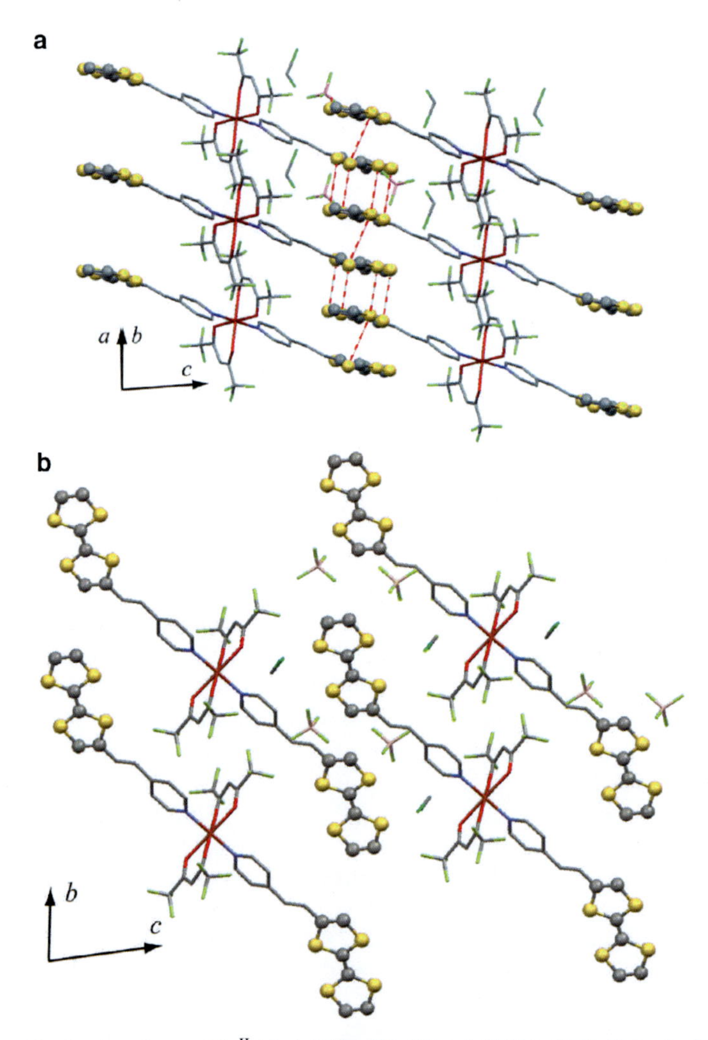

Fig. 9 (**a**) Projection of *trans*-$[Cu^{II}(hfac)_2(TTF\text{-}CH{=}CH\text{-}py)_2](BF_4)_2{\cdot}2CH_2Cl_2$ in the (120) plane showing the dimerized TTF chains and the segregation of BF_4^- from TTF chain. (**b**) Representation of the structure of *trans*-$[Cu^{II}(hfac)_2(TTF\text{-}CH{=}CH\text{-}py)_2](BF_4)_2{\cdot}2CH_2Cl_2$ showing alternating organic/inorganic networks and the BF_4^- anions separated from TTF layer

$(TTF{-}CH{=}CH{-}py)_2](PF_6){\cdot}2CH_2Cl_2$ was the first radical cation salt of a paramagnetic transition metal complex containing TTF−CH=CH−py as ligand.

The crystal structure consists of TTF layers in the (002) planes which alternate with inorganic $Cu(hfac)_2$ layers in the (001) planes. These two layers are linked by conjugated bridge (Fig. 10a). Figure 10b focuses on the organic plane; the TTF units form dimers with the shortest S···S contact equal to 3.593(3) Å. In one dimer, each TTF belongs to two different complexes with copper ions lying in two

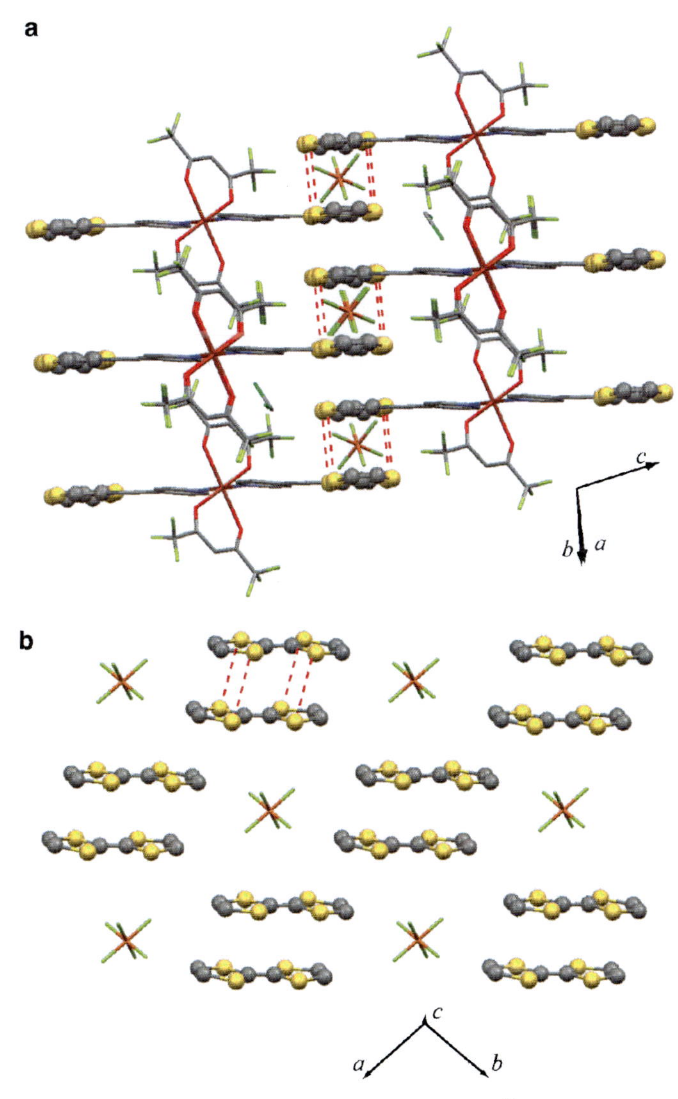

Fig. 10 **(a)** Structure of the mixed valence compound $[Cu^{II}(hfac)_2(TTF{-}CH{=}CH{-}py)_2](PF_6)\cdot 2CH_2Cl_2$, showing alternating organic/inorganic layers with the PF_6^- anions incorporated in the TTF layer. **(b)** Projection in the *ab* plane showing the separation of the TTF dimers by the PF_6^- anions

different inorganic layers. All dimers are separated by PF_6^- anions, and form mixed layers.

At 4 K the EPR spectrum of powdered sample shows signals typical of Cu^{II} with tetragonally elongated octahedral geometry, but no signal from the organic $TTF{-}CH{=}CH{-}py^{+\bullet}$ radical, which is only observable in a 10^{-3} M acetonitrile solution. The $A_{//}$ and $g_{//}$ values remain invariant up to 70 K, where they then begin

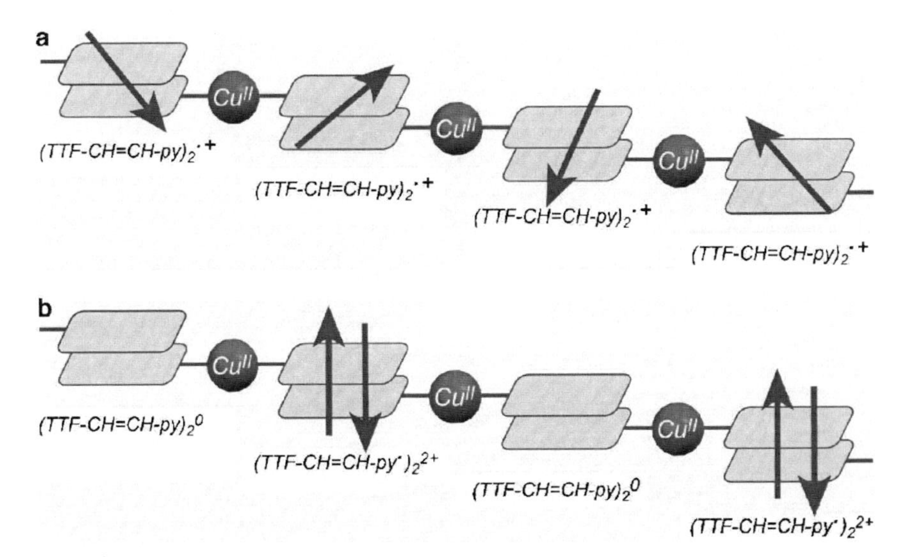

Fig. 11 Schematic crystal packing of $[Cu^{II}(hfac)_2(TTF-CH=CH-py)_2]^+$ for (**a**) homogeneous charge distribution and (**b**) charge disproportionated state

to decrease more rapidly on increasing temperature. Along with the nearly isotropic spectrum with g-value of about 2.15 at room temperature, these features are attributed to the dynamic Jahn-Teller distortion above 70 K.

Between 1.9 and 300 K its magnetic susceptibility χ can be fitted by the Curie-Weiss expression, $\chi \propto (T - \theta)^{-1}$, with Weiss temperature $\theta = -3.8$ K. The effective moment μ_{eff} value at room-temperature $(1.84(3)\mu_B)$ was found between the value expected from an uncorrelated two $S = 1/2$ spins composed of Cu^{II} and $(TTF-CH=CH-py)_2{}^{+\bullet}$ spins $(2.45\mu_B)$ and those expected for an independent $g = 2$, $S = 1/2$ system $(1.73\mu_B)$. The most probable reason for the absence of $(TTF-CH=CH-py)_2{}^{+\bullet}$ spins is the charge disproportionation on TTF−CH=CH−py dimers (Fig. 11b) instead of the homogeneous charged distribution (Fig. 11a).

On one hand, the partially oxidized TTFs are crucial to obtain conducting network but, on the other hand, the penetration of anions in the organic lattice prevent any significant electronic conductivity. The material is an insulator.

3 Polynuclear Compounds

3.1 *Binuclear Complex $Co^{II}_2(PhCOO)_4(Me_3TTF-CH=CH-py)_2$*

A larger trimethyl TTF−CH=CH−py derivative $(Me_3TTF-CH=CH-py)$ was coordinated to a Co^{II} benzoate dimer, having paddlewheel core structure [51].

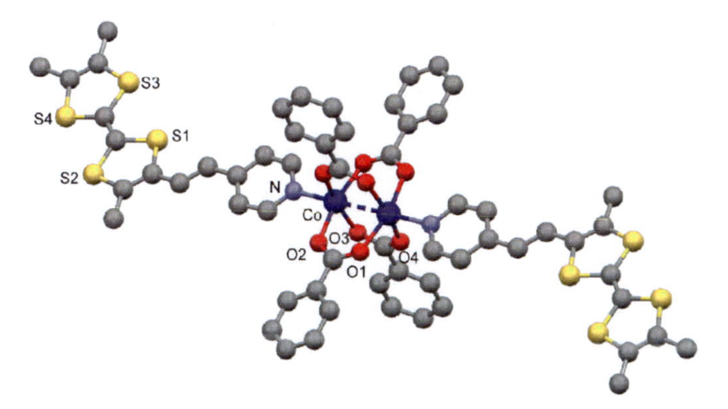

Fig. 12 Drawing of the paddlewheel complex $Co_2(PhCOO)_4(Me_3TTF{-}CH{=}CH{-}py)_2$

The complex was synthesized using the so-called in situ benzoate ligand generation from benzaldehyde [62]. The centrosymmetrical complex, formulated $Co^{II}{}_2(PhCOO)_4(Me_3TTF{-}CH{=}CH{-}py)_2$, is shown in Fig. 12.

The metal lies in a distorted square-pyramidal environment formed by four oxygen atoms of four benzoate ligands while the apical position is occupied by the nitrogen atom of the pyridyl ring of the TTF moiety. Four single benzoate $\mu_2(\eta_1,\eta_1)$ ions are coordinated to the cobalt dimer to give the paddlewheel core. One of the benzoates lies onto the mean plane of $Me_3TTF{-}CH{=}CH{-}py$ and is slightly closer to Co^{II} (Co–O distance shorter than 2.02 Å at room temperature) than the others (Co–O distances longer than 2.04 Å). The Co–N bond length is equal to 2.055(4) Å. Like other compounds presented here, the crystal packing consists of alternate organic and inorganic layers with intermolecular S···S distances longer than the sum of the vdw radii. Due to the ligand neutrality the compound is an insulator. On cooling, $\chi_M T$ decreases continuously down to zero. The ground state is therefore nonmagnetic while the observed paramagnetism above 70 K clearly indicates that states with single electrons are located close enough in energy from the ground state to be thermally populated. Boyd and coworkers [63, 64] demonstrated the importance of the through-bridge superexchange pathway in paddlewheel complexes. However, the magnetism of this type of dimer can be interpreted in terms of weak metal-metal interaction between the two Co^{II} ions. When considering an idealized D_{4h} symmetry for the dimer and the distance between Co^{II} ions (2.694(2) Å at 100 K), *d* orbitals of each Co^{II} with a *z* component overlap significantly (Fig. 13). The ground state of the dimer with the electronic configuration $\delta^2\delta^{*2}\pi^4\pi^{*4}\sigma^2$ is formally diamagnetic ($S = 0$) while the first excited state $\delta^2\delta^{*2}\pi^4\pi^{*4}\sigma^1\sigma^{*1}$ is paramagnetic ($S = 1$). Furthermore, if our assumption is true, the thermal population of the $S = 1$ state should destabilize metal–metal interaction. It is exactly what we observe with a significant increase of the metal–metal distance between 100 K and room temperature (3% increase while the increase of the other interatomic distances due to thermal expansion is only of the order of 1%).

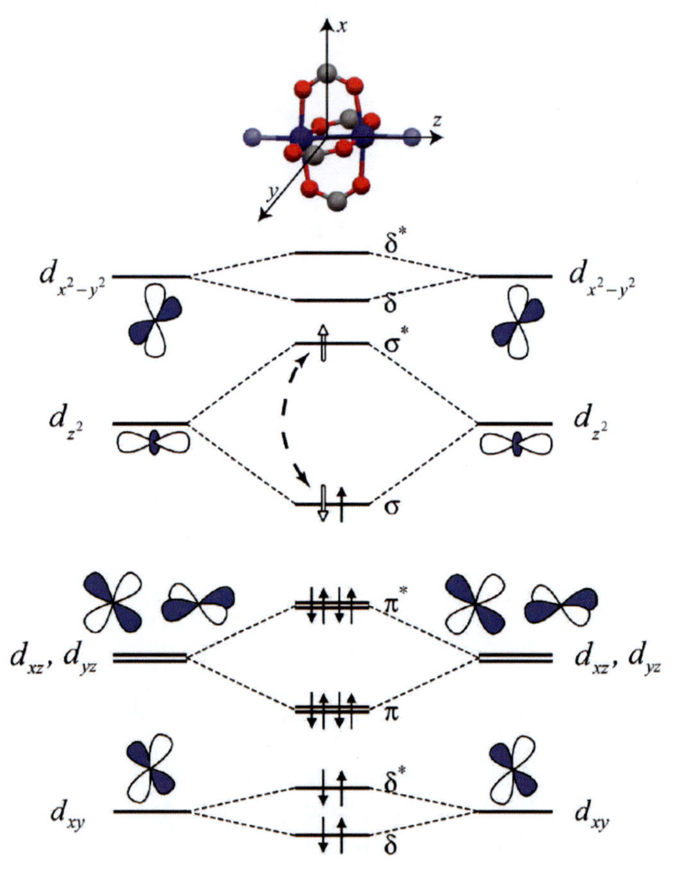

Fig. 13 Qualitative molecular orbital energy level diagram of the dimer *d* orbitals with the 14 electrons showing the electronic configuration $\delta^2\delta^{*2}\pi^4\pi^{*4}\sigma^2$ (singlet) and the first excited state $\delta^2\delta^{*2}\pi^4\pi^{*4}\sigma^1\sigma^{*1}$ (triplet)

3.2 Trinuclear Complexes $Co^{II}{}_2M^{II}(PhCOO)_6$ $(TTF-CH=CH-py)_2$ with M = Co, Mn

On our way to explore paramagnetic electroactive complexes of higher nuclearity, the TTF–CH=CH–py ligand was coordinated to the trinuclear core $[Co^{II}{}_2M^{II}(PhCOO)_6]$ [62]. Two isostructural electroactive paramagnetic centrosymmetric heterometallic (M = Mn) and homometallic (M = Co) complexes were synthesized (Fig. 14) [65]. In these compounds, the central metal ion (Mn^{II} or Co^{II}) occupies quite regular octahedron sites made of six oxygen atoms from different benzoates, four benzoates $\mu_2(\eta_1,\eta_1)$ bridging ligands, the last two benzoate being $\mu_2(\eta_1,\eta_2)$ bonded. The outer Co^{II} ions are in a distorted bipyramidal surrounding, designed with one nitrogen atom from the pyridine ring of the TTF–CH=CH–py ligand,

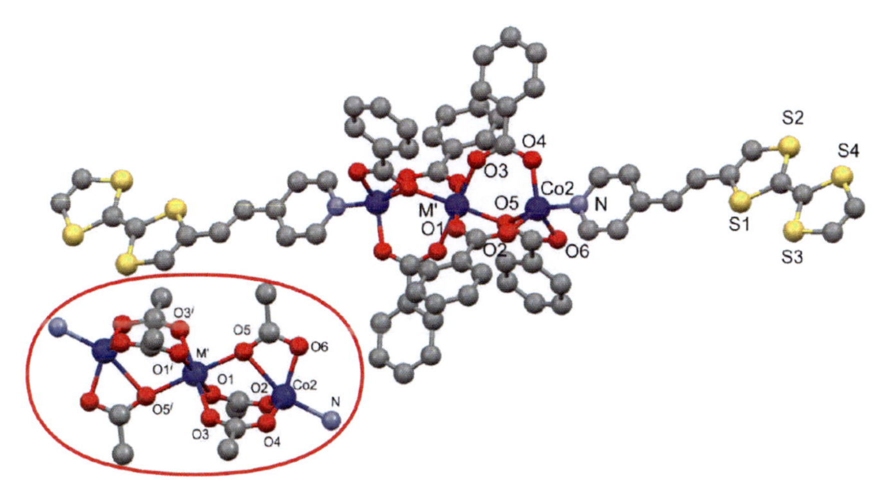

Fig. 14 Ball and stick representation of trinuclear complexes with M = Co or Mn (hydrogen atoms have been removed for simplicity). Coordination details are given separately (*in the circle*) where, for simplicity, we have removed aromatic rings of benzoates as well as TTF–CH=CH–py

three η_1 oxygen from three different benzoate anions and one η_2-O atom from the $\mu_2(\eta_1,\eta_2)$ benzoate ligand which is more distant from Co^{II}.

The bond lengths and angles of the TTF moiety are close to those reported for the noncoordinated neutral unit, indicating that organic molecule is neutral. The complexes reach the nanometric scale, the length of the molecules is equal to 38.25 Å (≈4 nm). The shortest intermolecular S···S contacts are equal to 3.9768(35) Å for homometallic complex and 3.9456(16) Å for heterometallic.

Despite the fact that the chemical formulation of the heterometallic complex is more complicated than that of the homometallic one, the magnetism is easily reproduced. The $\chi_M T$ vs T curve is almost completely flat down to 100 K with room temperature value equal to 9.57 cm^3 K mol^{-1} and obeys Curie law. The superexchange interactions between spin carriers are not strong enough to be detected at higher temperatures than 100 K. In the 2–100 K temperature range the magnetism can be reproduced with good agreement considering antiferromagnetic nearest-neighbor superexchange interactions. The Hamiltonian employed is

$$H = -J\boldsymbol{S}_{\mathrm{Co}}\,\boldsymbol{S}_{\mathrm{Mn}} + g_{\mathrm{Co}}\,\beta \boldsymbol{H}\,\boldsymbol{S}_{\mathrm{Co}} + g_{\mathrm{Mn}}\,\beta \boldsymbol{H}\,\boldsymbol{S}_{\mathrm{Mn}} \quad \text{with } \boldsymbol{S}_{\mathrm{Co}} = \boldsymbol{S}_{\mathrm{Co}_1} + \boldsymbol{S}_{\mathrm{Co}_2}.$$

In fixing g_{Mn} = 2.00 the best fit is obtained with $J = -1.80$ cm^{-1} and g_{Co} = 2.29 (Fig. 15). The ground state of the system corresponds to total spin $S = 1/2$ ($S = S_{Co} + S_{Mn}$) with Mn^{II} spin antiparallely aligned to the terminal Co^{II} spins. On cooling, $\chi_M T$ vs T for the homometallic species decreases from 8.56 cm^3 K mol^{-1} at room temperature down to 6.23 cm^3 K mol^{-1} at 7.5 K, then increases on cooling further. The decrease of $\chi_M T$ on lowering the temperature results from the combination of two phenomena: (1) at very low temperature, octahedral Co^{II} can be described by an

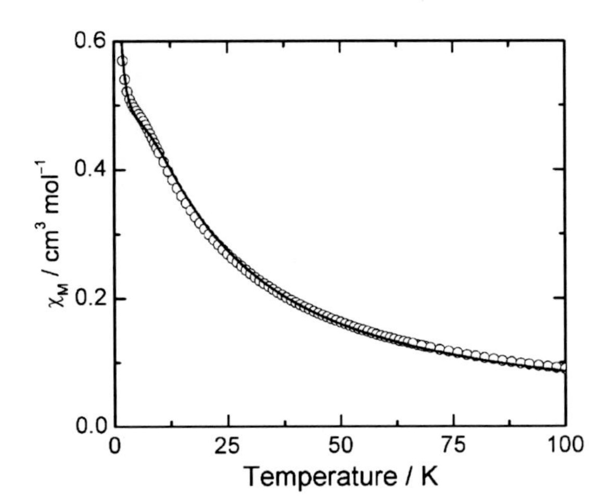

Fig. 15 Temperature dependence of the molar magnetic susceptibility (χ_M) of $Co^{II}_2Mn^{II}$ $(PhCOO)_6$ (TTF−CH=CH−py)$_2$ (*white circle*) with the best fitted curve (*full line*)

effective spin $S_{eff} = 1/2$ with an effective Zeeman factor as high as $g' = 4.3$ (due to local distortion and spin orbit coupling), and with large Temperature Independent Paramagnetism (TIP); (2) superexchange interaction between nearest neighbors. At low temperature, the system should fit with a ferrimagnetic trinuclear unit with $S_{eff} = 1/2$ antiferromagnetically coupled with the S_{Coi}s. Depending on the amplitude of the interaction, a minimum in $\chi_M T$ could be observed above 2 K. Since the amplitude of the experimental minimum is sample dependent, any attempts to fit the data in the low temperature regime (below 20 K) would be questionable.

4 Conclusion

We have demonstrated throughout this chapter that 3*d* transition metals can easily be coordinated by redox active ligands derived from TTF. Our aim was to design complexes with a covalently linked π electrons (itinerant electrons) of TTFs and *d* electrons (paramagnetic electrons) of transition metal ions. By using these chemical tools we have been able to produce several complexes with various nuclearities, from mononuclear up to trinuclear homo- or heterometallic species, and also with various metal:redox active ligand ratio, from 1:1 up to 3:2. We have presented materials in which the TTF moieties of the complexes are neutral, and materials in which the TTF are fully oxidized. Also, an interesting material containing mixed valence TTF-type organic donor ligand is presented.

So far, in all the materials investigated, the charge of the complexes follows the local charges of TTFs. For example, complexes are all neutral when TTF-based ligands are neutral and cationic when TTFs are partially or fully oxidized. Therefore, anions are required in the crystal lattice to counterbalance the TTFs charges, when they are in an oxidized state. Several types of packing have been characterized. One is made of mixed layers of TTFs/anions in which alternate organic donors and inorganic anions appear. In a second type of packing, there is segregation between inorganic and organic moieties with pure organic layers made of monocationic TTFs.

The possibility of coordinating functionalized TTFs onto polynuclear core is a very stimulating issue because it is now well established that polynuclear cores, with some restrictions of course, can act as SMMs. We started a systematic investigation of polynuclear paramagnetic complexes with TTF−CH=CH−py ligands to scan the possibility to access to bifunctional molecules which can act at the same time as SMM and single component metal. We succeeded in coordinating our modified TTFs to several homo- or heteropolynuclear complexes. This opens new perspectives in the field of multifunctional materials. The size of these molecules, which is of the order of 4 nm, is another important aspect in the field of molecular scale electronic.

References

1. Akamutu H, Inokuchi H, Matsunaga Y (1954) Nature 173:168
2. Ferraris J, Cowan DO, Walatka VJ, Perlstein JH (1973) J Am Chem Soc 95:948
3. Coleman LB, Cohen MJ, Sandman DJ, Yamagishi FG, Garito AF, Ferraris AF (1973) Solid State Commun 12:1125
4. Bechgaard K, Jacobsen CS, Mortensen K, Pedersen HJ, Thorup N (1980) Solid State Commun 33:1119
5. Bechgaard K, Carneiro K, Rasmussen FB, Olsen M, Rondorf G, Jacobsen CS, Pedersen HJ (1981) J Am Chem Soc 103:2440
6. Jérome D, Mazaud A, Ribault M, Bechgaard K (1980) J Phys Lett 41:L95
7. Miller JS, Calabrese C, Epstein AJ, Bigelow RW, Zang JH, Reiff WM (1986) J Chem Soc Chem Commun 1026
8. Pei Y, Verdaguer M, Kahn O, Sletten J, Renard JP (1986) J Am Chem Soc 108:428
9. Gatteschi D, Sessoli R (2003) Angew Chem Int Ed 42:2 and references therein
10. Caneschi A, Gatteschi D, Lalioti N, Sangregorio C, Sessoli R, Venturi G, Vindigni A, Rettori A, Pini MG, Novak MA (2001) Angew Chem Int Ed 40:1760
11. Clérac R, Miyasaka M, Yamashita M, Coulon C (2002) J Am Chem Soc 124:12837
12. Day P, Kurmoo M, Mallah T, Marsden IR, Friend RH, Pratt FL, Hayes W, Chasseau D, Gaultier J, Bravic G, Ducasse L (1992) J Am Chem Soc 114:10722
13. Graham AW, Kurmoo M, Day P (1995) J Chem Soc Chem Commun 2061
14. Kobayashi H, Tomita H, Naito T, Kobayashi A, Sakai F, Watanabe T, Cassoux P (1996) J Am Chem Soc 118:368
15. Martin L, Turner SS, Day P, Mabbs FE, McInnes EJL (1997) Chem Commun 1396
16. Zhang B, Wang Z, Zhang Y, Takahashi K, Okano Y, Cui H, Kobayashi H, Inoue K, Kurmoo M, Pratt FL, Zhu D (2006) Inorg Chem 45:3275
17. Coronado E, Galán-Mascarós JR, Gómez-García CJ, Laukhin V (2000) Nature 408:447
18. See for example the special issue: (2004) Chem Rev 11
19. Ouahab L, Enoki T (2004) Eur J Inorg Chem 933
20. Kobayashi H, Kobayashi A, Cassoux P (2000) Chem Soc Rev 29:325
21. Fujiwara H, Fujiwara E, Nakazawa Y, Narymbetov BZ, Kato K, Kobayashi H, Kobayashi A, Tokumoto M, Cassoux P (2001) J Am Chem Soc 123:306

22. Uji S, Shinagawa H, Terashima T, Yakabe T, Terai Y, Tokumoto M, Kobayashi A, Tanaka H, Kobayashi H (2001) Nature 410:908
23. Zhang B, Tanaka H, Fujiwara H, Kobayashi H, Fujiwara E, Kobayashi A (2002) J Am Chem Soc 124:9982
24. Coronado E, Georges R, Tsukerblat B (1996) In: Coronado E, Delhaès P, Gatteschi D, Miller JS (eds) Molecular magnetism: from molecular assemblies to the devices, vol E321 (NATO ASI Series). Kluwer, The Netherlands, p 65
25. Imakubo T, Sawa H, Kato R (1995) J Chem Soc Chem Commun 1667
26. Imakubo T, Sawa H, Kato R (1996) Mol Cryst Liq Cryst 285:27
27. Imakubo T, Sawa H, Kato R (1997) Synth Met 86:1883
28. Imakubo T, Sawa H, Kato R (1995) Synth Met 73:117
29. Imakubo T, Tajima N, Tamura M, Kato R (2002) J Mater Chem 12:159
30. Thoyon D, Okabe K, Imakubo T, Golhen S, Miyazaki A, Enoki T, Ouahab L (2002) Mol Cryst Liq Cryst 376:25
31. Hervé K, Cador O, Golhen S, Costuas K, Halet JF, Shirahata T, Muto T, Imakubo T, Miyazaki A, Ouahab L (2006) Chem Mater 18:790
32. ADF2004.01 (2003) Theoretical chemistry. Vrije Universiteit, SCM, Amsterdam, the Netherlands, http://www.scm.com
33. Baerends EJ, Ellis DE, Ros P (1973) Chem Phys 2:41
34. te Velde G, Bickelhaupt FM, Fonseca Guerra C, van Gisbergen SJA, Baerends EJ, Snijders JG, Ziegler T (2001) J Comput Chem 22:931
35. Zhang B, Wang ZM, Fujiwara H, Kobayashi H, Kurmoo M, Inoue K, Mori T, Gao S, Zhang Y, Zhu DB (2005) Adv Mater 17:1988
36. Turner SS, Le Pévelen D, Day P (2003) Synth Met 133/134:497
37. Setifi F, Golhen S, Ouahab L, Miyazaki A, Okabe K, Enoki T, Toita T, Yamada JI (2002) Inorg Chem 41:3786
38. Setifi F, Ouahab L, Golhen S, Miyazaki A, Enoki T, Yamada JI (2003) C R Chimie 6:309
39. Hiraga H, Miyasaka H, Nakata K, Kajiwara T, Takaishi S, Oshima Y, Nojiri H, Yamashita M (2007) Inorg Chem 46:9661
40. Yamada J, Sugimoto T (2004) TTF chemistry. Springer, Berlin Heidelberg New York
41. Wu JC, Liu SX, Keene TD, Neels A, Mereacre V, Powell AK, Decurtins S (2008) Inorg Chem 47:3452
42. Ota A, Ouahab L, Golhen S, Cador O, Yoshida Y, Saito G (2005) New J Chem 29:1135
43. Liu SX, Ambrus C, Dolder S, Neels A, Decurtins S (2006) Inorg Chem 45:9622
44. Lu W, Zhang Y, Dai J, Zhu QY, Bian GQ, Zhang DQ (2006) Eur J Inorg Chem 1629
45. Setifi F, Ouahab L, Golhen S, Yoshida Y, Saito G (2003) Inorg Chem 42:1791
46. Jia C, Liu SX, Ambrus C, Neels A, Labat G, Decurtins S (2006) Inorg Chem 45:3152
47. Liu SX, Dolder S, Franz P, Neels A, Stoeckli-Evans H, Decurtins S (2003) Inorg Chem 42:4801
48. Iwahori F, Golhen S, Ouahab L, Carlier R, Sutter JP (2001) Inorg Chem 40:6541
49. Ouahab L, Iwahori F, Golhen S, Carlier R, Sutter JP (2003) Synth Met 133/134:505
50. Hervé K, Liu SX, Cador O, Golhen S, Le Gal Y, Bousseksou A, Stoeckli-Evans H, Decurtins S, Ouahab L (2006) Eur J Inorg Chem:3498
51. Benbellat N, Gavrilenko KS, Le Gal Y, Cador O, Golhen S, Gouasmia A, Fabre JM, Ouahab L (2006) Inorg Chem 45:10440
52. Gavrilenko KS, Le Gal Y, Cador O, Golhen S, Ouahab L (2007) Chem Commun 280
53. Mosimann M, Liu SX, Labat G, Neels A, Decurtins S (2007) Inorg Chim Acta 360:3848
54. Xu W, Zhang D, Li H, Zhu D (1999) J Mater Chem 9:1245
55. Massue J, Bellec N, Chopin S, Levillain E, Roisnel T, Clérac R, Lorcy D (2005) Inorg Chem 44:8740
56. Bellec N, Massue J, Roisnel T, Lorcy D (2007) Inorg Chem Commun 10:1172
57. Zhu QY, Bian GQ, Zhang Y, Dai J, Zhang DQ, Lu W (2006) Inorg Chim Acta 359:2303

58. Chahma M, Hassan N, Alberola A, Stoeckli-Evans H, Pilkington M (2007) Inorg Chem 46:3807
59. Andreu R, Malfant I, Lacroix PG, Cassoux P (2000) Eur J Org Chem 737
60. Setifi F (2003) PhD thesis, Université de Rennes 1, France
61. Hervé K, Le Gal Y, Ouahab L, Golhen S, Cador O (2005) Synth Met 153:461
62. Gavrilenko KS, Punin SV, Cador O, Golhen S, Ouahab L, Pavlishchuk VV (2005) J Am Chem Soc 127:12246
63. Boyd PDW, Gerloch M, Harding JH, Wolley RG (1978) Proc R Soc Lond A360:161
64. Boyd PDW, Davies JE, Gerloch M (1978) Proc R Soc Lond A360:191
65. Gavrilenko KS, Le Gal Y, Cador O, Golhen S, Ouahab L (2007) Chem Commun 280

Top Organomet Chem (2009) 27: 77–96

π–*d* Interaction-Based Molecular Conducting Magnets

Akira Miyazaki and Toshiaki Enoki

Abstract The crystal structures and electronic and magnetic properties of conducting molecular magnets developed by our group are reviewed, from the viewpoints of our two strategies for increasing the interaction between the conducting π-layer and magnetic *d*-layer (π–*d* interaction). $(DMET)_2FeBr_4$ and $(EDTDM)_2FeBr_4$ is composed of quasi-one-dimensional donor sheets sandwiched between magnetic anion sheets. The ground state of the donor layer changes from the insulator state to the metallic state by the application of pressure especially for the EDTDM salt, and near the insulator-metal phase boundary pressure the magnetic order of the anion spins considerably affects the transport properties of the donor layer. The crystal structure of $(EDT\text{-}TTFBr_2)_2FeBr_4$ and $(EDO\text{-}TTFBr_2)_2FeX_4$ (X = Cl, Br) is characterized by strong intermolecular halogen-halogen contacts between the organic donor and FeX_4^- anion molecules. The magnetic order of the anion spins and relatively high magnetic order transition temperature proves the effectiveness of the halogen-halogen contacts to strengthen the π-*d* interaction.

Keywords π–*d* Interaction, Molecular conductors, Molecular magnets

Contents

A. Miyazaki(✉) and T. Enoki

Department of Environmental Applied Chemistry, University of Toyama, 3190 Gofuku, Toyama-shi, Toyama, 930-8551, Japan, E-mail: miyazaki@eng.u-toyama.ac.jp

Department of Chemistry, Tokyo Institute of Technology, 2-12-1-W4-1 Ookayama, Meguro-ku, Tokyo 152-8551, Japan

M. Fourmigué and L. Ouahab (eds.), *Conducting and Magnetic Organometallic Molecular Materials*, Topics in Organometallic Chemistry 27,
DOI: 10.1007/978-3-642-00408-7_4,

1 Introduction

Physical properties of molecular metals and superconductors are one of the principal areas of interest in the fields of solid-state physics and chemistry [1–3]. The history of molecular conductors dates back to the pioneering works in the 1950s on condensed aromatic hydrocarbon compounds such as perylene or violanthrene doped with halogens [4, 5]. These compounds can be regarded as ancestors of graphene (single-layer graphite) which has recently drawn attention with its unusual electronic states [6]. In the 1970s, heterocyclic aromatic compounds such as TTF (tetrathiafulvalene, Fig. 1) and its derivatives were regarded as a promising class of organic donor molecules, because of their great ability to produce metallic electronic states either in charge-transfer complexes or radical ion salts. For example, a parent molecule TTF forms a charge-transfer complex with TCNQ (tetracyanoquinodimethane) [7] that shows a metallic conductivity that is comparable or surpasses the conductivity of copper metal around liquid nitrogen temperatures, although the metallic state is suddenly lost below 54 K due to the structural phase transition [8]. The series of organic superconductors $TMTSF_2X$ (TMTSF = tetramethyltetraselenafulvalene, X = PF_6, ClO_4, etc.) [9, 10] continuously drew the attention especially of physicists for about 30 years after their discovery in the early 1980s because of their richness in unconventional physical properties such as charge and spin density wave states [11], charge-ordered states

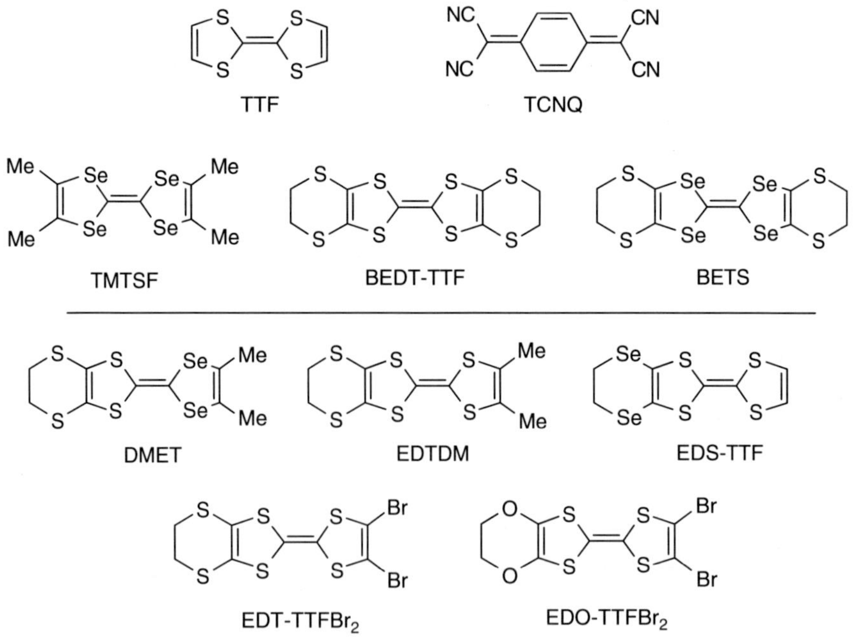

Fig. 1 Molecular structure of organic donor molecules that appear in this chapter

[12, 13], Tomonaga–Luttinger liquid state [14–16], unconventional superconductivity [17], etc. These molecular conductors once used to be called "organic metals," but nowadays this terminology has become obsolete in order to avoid possible confusion with "organometallics."

Compared to inorganic atom-based materials, molecular conductors have the following two distinguishing characteristics. The first is the low-dimensionality in their electronic nature. TTF derivative molecules are planar in general and their molecular orbitals are made of π-orbitals, hence the electronic systems of the molecular conductors are highly anisotropic and bear one- or two-dimensional character. Due to this low-dimensionality, molecular materials exhibit a large variety of electronic ground states as mentioned above. Since many of them are insulator phases, this low-dimensionality at first glance would seem to be disadvantageous. However, this variety of electronic states due to the low-dimensionality is the attractive point of molecular conductors, since these different electronic states are often adjacent to each other in the phase diagrams and can easily be transferred from one to another by small external stimuli such as pressure or magnetic field. A recently discovered organic thyristor [18] is also based on this characteristic; the mechanism is associated with the melting of a charge-ordered state to a metallic state by applying an external current.

The second characteristic is a high potential for chemical modification of the constituent molecules, which strongly relies on the activities of synthetic organic chemistry [19]. The TTF molecules contain four vinyl hydrogen atoms in two five-membered 1,3-dithiole rings, which can easily be substituted with other functional groups using relatively simple synthetic procedures. This versatility enlarges the field of the novel electronic system based on the family of TTF derivatives. For example, the BEDT-TTF molecule produces a large number of metals and superconductors, such as β'-(BEDT-TTF)$_2$ICl$_2$ with the highest superconducting transition temperature among TTF-based molecular conductors (14.2 K at 8.2 GPa) [20].

The hybridization of such molecular metals with magnetic transition metal complexes presents an important feature to physics and chemistry. If magnetic ions are introduced as a counterpart for organic donors as will be discussed in the next section, novel molecular materials having cooperative transport and magnetic properties are anticipated. We have developed a number of molecular magnets based on TTF-type derivatives and magnetic anions, and investigated their physical properties in detail [21–35], some of which have been summarized in our preceding review articles [36–39]. Such conducting molecular magnets have also attracted many research groups to develop the field of molecule-based magnets [40–44]. One of the representative examples is λ-(BETS)$_2$FeCl$_4$ (BETS = bis(ethylenedithio) tetraselenafulvalene) [45, 46], in which an application of magnetic field replaces a magnetic insulating state with a field-induced superconductive phase.

In the next section we discuss and summarize our guidelines to construct molecular conducting magnets by means of the exchange interaction between the conducting π-layer and magnetic d-layer (π–d interaction). The following two sections then introduce several examples of π–d interaction based conducting molecular magnets developed by our group.

2 Recipes for Conducting Molecular Magnets

For the rational design of molecular conducting magnets, it is useful to remind ourselves why conventional, atom-based magnetic metals, such as Fe or rare-earths, become magnets. These elemental magnets are divided into two groups according to the mechanism of their magnetism. The first group consists of 3*d* transition metals like Fe, Co and Ni, of which magnetism can be understood using the Stoner model [47]. The conduction electrons of these metals are a hybridization of 3*d*, 4*s* and 4*p* electrons, and thus these metals have a complicated electronic band structure having a sharp peak which schematically illustrated in Fig. 2a. Due to the Coulomb interaction between conduction electrons and Pauli exclusion principle, the energy bands of up- and down-spinning electrons are oppositely shifted (exchange splitting), resulting in the nonvanishing magnetic moment. Unfortunately, this Stoner model strongly relies on the shape and nature of the electronic band structure, which is in fact difficult to design and predict from the molecular approach.

The other group contains rare earths such as lanthanides, in which 6*s* electrons behave as itinerant conduction electrons and 4*f* electrons are localized at their atomic position to produce localized moments. Alloys containing a small amount of magnetic 3*d* transition metal, such as Co-Pd alloy, can also be included in this category. The spin alignment mechanism of these materials is well described with the RKKY (Ruderman–Kittel–Kasuya–Yosida) [48–50] mechanism. A localized magnetic moment causes the spin density oscillation in conduction electrons through *s–f* or *s–d* interaction as shown in Fig. 2b, which then interacts with other localized spins to establish magnetic exchange interaction. In contrast to the Stoner magnet, this RKKY magnet can be designed using molecular materials; the combination of the itinerant and localized electron systems can be constructed using organic donor molecules and transition metal complexes, respectively. Here the interaction between the itinerant π-electrons of the organic part and localized *d*-electrons in the inorganic part is referred as the π–*d* exchange interaction.

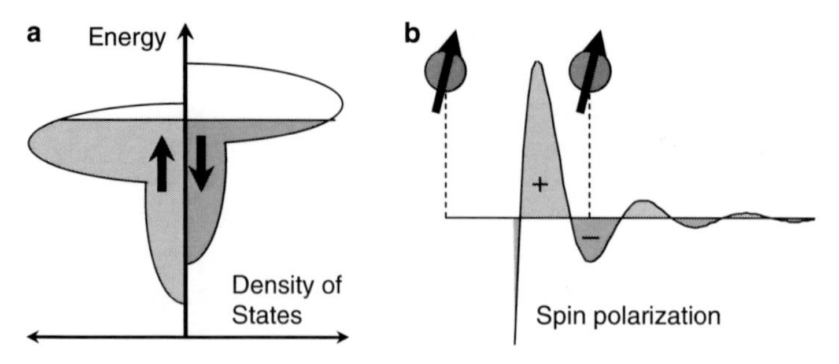

Fig. 2 Schematic diagrams of two representative mechanisms for spin-alignment in conducting magnets. (**a**) Stoner mechanism. (**b**) RKKY mechanism

The electronic ground state of the π-electron system in molecular conductors ranges from metallic state to insulator as mentioned in the last section. When a metal-insulator transition takes place in the π-electron system, the Stoner magnets fall into nonmagnetic insulator, since the presence of the delocalized electrons are indispensable for this mechanism. For the RKKY magnets, in contrast, even if the organic π-electron systems have insulating ground states due to the low-dimensional instability and/or strong correlation between the π-electrons (Mott insulator, SDW), the localized π-electrons can couple with the magnetic moments of *d*-electrons through the π–*d* exchange interaction to produce composite magnets.

Obviously the π–*d* exchange interaction plays a principal role in these molecule-based conducting magnets. Before describing our strategy for achieving interaction, we first review the discussion on the electronic structure of the organic conductors in terms of the Mott-Hubbard Hamiltonian because of its similarity with the mechanism of the π-*d* exchange interaction. This model Hamiltonian is characterized with two parameters, the transfer integral t and the on-site Coulomb repulsion energy U. The transfer integral t is expressed as

$$t = \int \psi_i^*(r)\hat{H}\psi_{i+1}(r)\mathrm{d}r, \tag{1}$$

where $\hat{\mathrm{H}}$ is the one-electron Hamiltonian of the π-electrons and $\psi_i\,(r)$ is the wave function of HOMO (highest occupied molecular orbital) of the i-th molecule. In the Hückel molecular orbital theory this transfer integral t is equivalent to the resonance integral β. In a crude approximation, the integral t is proportional to the overlap integral

$$S = \int \psi_i^*(r)\psi_{i+1}(r)\mathrm{d}r \tag{2}$$

between the molecular orbitals, which is easier to be evaluated from the molecular and crystal structure.

The on-site Coulomb repulsion energy U is defined as

$$U = E(2) + E(0) - 2E(1), \tag{3}$$

where $E(n)$ is the energy of a donor molecule whose HOMO contains n electrons. The quantity U can be understood as the Coulomb repulsive interaction between two electrons in a single molecular orbital, since under (extended) Hückel approximation without electron-electron interaction, $E(2)+E(0)$ is equal to $2E(1)$ and U equals to zero. From the molecular approach, U can roughly be evaluated as

$$U \approx e^2/\varepsilon r, \tag{4}$$

where ε is the dielectric constant and r is the size of the π-electron system on the molecule.

Using these parameters, the electronic state of molecular conductors is described with the Hubbard–Hamiltonian

$$\hat{H} = \sum_i U_i n_{i\uparrow} n_{i\downarrow} + \sum_{i,\sigma} t_i \left(c_{i,\sigma}^{+} c_{i-1,\sigma} + c_{i-1,\sigma}^{+} c_{i,\sigma} \right), \tag{5}$$

where $c_{i,\sigma}^{+}$ and $c_{i,\sigma}$ are the creation and annihilation operators, respectively, for the electron on the i-th molecule having σ (= $\uparrow$ or $\downarrow$) spin, and $n_{i,\sigma} \equiv c_{i,\sigma}^{+} c_{i,\sigma}$ is the number operator for this electron. The second term of the Hamiltonian is equivalent to the Hückel Hamiltonian, and describes the kinetic energy of an electron which jumps from molecule $i-1$ to molecule i and vice versa. Since the integrals t_i take negative values, this movement energetically stabilizes the π-electron system. The first term appears only if two electrons having opposite spins present on the same molecule and suppresses the intermolecular motion of the electrons because U is positive.

The competition between these two terms produces a large variety of electronic structures in molecular systems. The condition $t \gg U$ favors itinerant metallic states, whereas the condition $t \ll U$ stabilizes localized insulating states. In the latter case, the Hubbard Hamiltonian is reduced to the Heisenberg Hamiltonian

$$\hat{H} = -J \sum_i S_i S_{i+1}, \tag{6}$$

where $J = -4t^2/U$ is the exchange interaction and S_i is the spin operator for the electron located on the i-th molecule.

This discussion is also applicable to the π–d interacting system with slight modification. The Hubbard Hamiltonian for a pair of π-electron donor and d-electron magnetic anion is expressed as

$$\hat{H} = U_\pi n_{\pi\uparrow} n_{\pi\downarrow} + U_d n_{d\uparrow} n_{d\downarrow} + t_{\pi d} \sum_\sigma \left(c_{\pi\sigma}^{+} c_{d\sigma} + c_{d\sigma}^{+} c_{\pi\sigma} \right), \tag{7}$$

where suffixes π and d signify the π- and d-electron parts, respectively, and $t_{\pi d}$ is the transfer integral between them. Solving this Hamiltonian, the magnitude of the π–d exchange interaction is evaluated as

$$J = -2t_{\pi d}^2 \left(\frac{1}{U_\pi} + \frac{1}{U_d} \right). \tag{8}$$

This result theoretically rationalizes our guidelines for achieving large π–d exchange interaction: (1) the on-site Coulomb repulsions for both donor and anion molecules should be reduced, and (2) the transfer integral between them should be increased. Iron tetrahalides FeX_4^- (X = Cl, Br) satisfy both of these requirements and they are frequently used as magnetic anion systems. In fact, the unpaired d-electrons of the central Fe metal are delocalized to the Cl or Br ligands due to the partial covalent character between the metal and ligand, which is experimentally proved for a uniaxially-distorted $FeCl_4^-$ anion showing a single-ion magnetic anisotropy [51]. Due to this delocalization, the on-site Coulomb repulsion U_d is reduced in these magnetic anions. This delocalization at the same time increases the spin density on the ligands that are located close to π-electron donor counterparts, and the absolute value of intermolecular transfer integral $|t_{\pi d}|$ is also enhanced.

The FeX_4^- ions also have several practical advantages. These anions are thermodynamically and kinetically stable monovalent anions, and thus radical ion salts are ready to be obtained by conventional electrocrystallization technique. They also have large spin quantum number $S = 5/2$ and the magnetism of the d-electron system is easy to investigate. In particularl, the FeX_4^- ions have non-magnetic analogues of the anion, GaX_4^-, which often give radical ion salts of isostructural crystal due to the similarity in ionic radii of Ga^{3+} and Fe^{3+} [52]. This nonmagnetic analogue is useful in comprehensively investigating the effect of π–d interaction; the nature of the π-electron system can be investigated using the GaX_4 salt, and the interplay between the π- and d-electrons will be clarified by the comparison of the results for FeX_4 and GaX_4 salts. We can even tune and modify the magnitude of the π–d interaction by making $Fe_xGa_{1-x}X_4$ or $FeCl_yBr_{4-y}$ alloys in the counter anion layer [45]. The correlation between the chemical composition and the physical properties gives deep insight on the nature of the π–d interaction.

3 Conducting Magnets Having Quasi One-Dimensional Electron System

In this and the next sections we discuss two groups of molecule-based conducting magnets at which the π–d interaction works effectively. The first approach is the use of quasi one-dimensional electronic systems as the π-electron layers, and the other strategy is to increase the magnitude of the π–d interaction by the introduction of intermolecular halogen-halogen contacts.

As pointed out previously, quasi one-dimensional electronic systems tend to undergo metal-insulator transitions due to their electronic instability. Although their electronic ground states are difficult to predict in general, one typical case is a spin density wave (SDW) state caused by the electron correlation. This SDW state has a fractional magnetic moment and can easily be converted into the metallic state by the application of low external pressure. In the marginal region of these two phases, the π-electron system is expected to be sensitive to small external stimuli such as the spin ordering of the d-electron system.

Here we discuss interplay between the electron transport of the π-electrons and the magnetism of d-electron spins for three salts, $(DMET)_2FeBr_4$ [26], $(EDTDM)_2FeBr_4$ [23] and $(EDS\text{-}TTF)_2FeBr_4$ [53]. The crystal structures of these

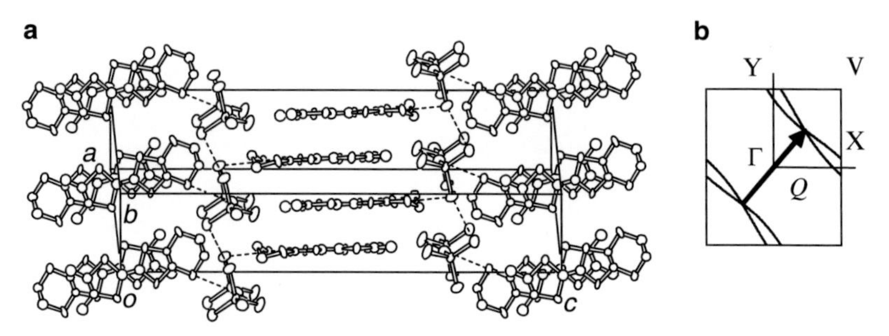

Fig. 3 (**a**) Crystal structure of $(DMET)_2FeBr_4$. The *dotted* and *dashed lines* denote the intermolecular anion—anion and donor–anion contacts, respectively. (**b**) Fermi surfaces obtained for a donor layer around $z = 1/2$ using the tight-binding approximation. The *solid arrow* represents the nesting vector $\boldsymbol{Q} \sim (\boldsymbol{a}^* \pm \boldsymbol{b}^*)/2$

salts are close to each other regardless of the donor molecules (Fig. 3). The structure is characterized by an alternating stack of donor and anion layers along the c-axis. In the donor layers, the organic donor molecules construct slightly dimerized columns elongated along the $a + b$ and $a - b$ directions according to the layers. As the transfer integrals inside the columns are about 10 times larger than those between the chains, the donor layers are characterized as a quasi one-dimensional system. As a result, the Fermi surfaces obtained by the tight-binding band calculation have wavy shapes with a possible nesting vector $\boldsymbol{Q} \sim (\boldsymbol{a}^* \pm \boldsymbol{b}^*)/2$ as shown in Fig. 3b. The tetrahedral $FeBr_4^-$ anions form a distorted square lattice formed by Br···Br contacts, although the Br···Br distances are longer than the van der Waals distance [54]. Between the donor columns and magnetic anion close S···Br or Se···Br contacts are observed. It should also be noted that for these donors, $GaBr_4$ salts with same crystal structure as corresponding $FeBr_4$ salts were also obtained.

The electrical conductivities of the $(DMET)_2FeBr_4$ and $(EDTDM)_2FeBr_4$ are metallic down to $T_{MI} \sim 40$ and 15 K, respectively, at which metal-insulator transitions are observed. The low temperature insulator phase is characterized as an SDW state, since the ESR signals of the donor cation radical broaden below T_{MI}, indicating the presence of magnetic order in the donor layer. From the temperature dependence of the susceptibility of isomorphous $(EDTDM)_2GaBr_4$ with nonmagnetic counter anion, the magnetic moment in the donor chain is estimated as 0.29 μ_B per donor molecule. On the other hand, $(EDS\text{-}TTF)_2FeBr_4$ undergoes a gradual metal-insulator transition at ca. 250 K, presumably due to the stronger one-dimensional character in the π-electron layer.

The magnetic susceptibilities of these three salts obey the Curie–Weiss law in high temperature range. The Curie constant C (4.4–4.7 emu K^{-1} mol^{-1}) is in good agreement with the calculated value of the $S = 5/2$ magnetic anion contribution. For the DMET and EDTDM salts, antiferromagnetic transitions are observed at $T_N = 3.7$ K and 3.0 K, respectively. The magnetization curve at 1.8 K (Fig. 4a) shows a spin-flop transition at B_{SF} when the external field is applied parallel to the a axis. The easy-axis magnetization curve is beyond the hard-axis magnetization

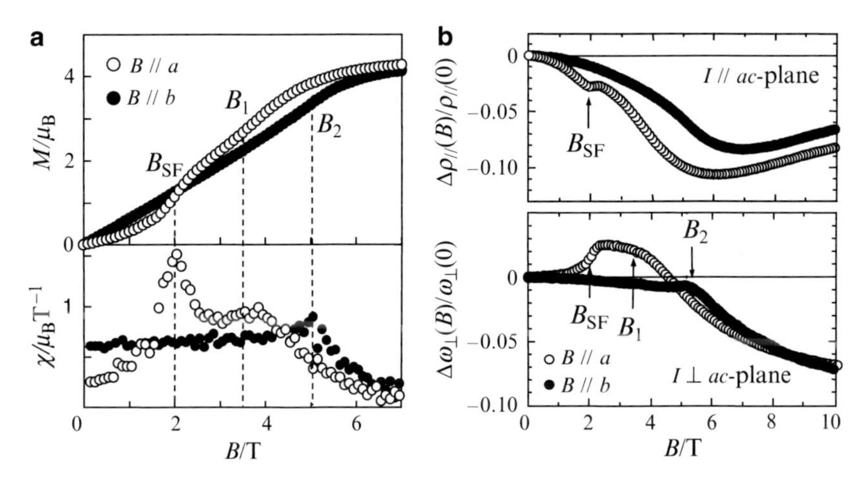

Fig. 4 (**a**) Magnetization curves of $(DMET)_2FeBr_4$ at 1.8 K. Magnetic susceptibilities ($\chi = dM/dB$) are also plotted as a function of the field. (**b**) Field dependence of the magnetoresistance for the in-plane transport $\Delta\rho_{//}(B)/\rho_{//}(0)$ and the interplane hopping frequency $\Delta\omega_{\perp}(B)/\omega_{\perp}(0)$ at 1.6 K. B_{SF} denotes the spin-flop transition and B_1 and B_2 show anomalies

curve above the spin flop field B_{SF}, and has a shoulder around B_1. A similar shoulder is also observed for the b-axis magnetization curve around B_2. When the donor molecule is replaced from DMET into EDTDM, the spin-flop field B_{SF} is almost unchanged whereas the anomalies at B_1 and B_2 appear at the lower field, suggesting that the origin of the latter two anomalies are related to the π-electron system. For the EDS-TTF salt, despite the presence of the close Br···Se contacts between the anion and donor layers, the absence of magnetic transition and the negligible Weiss temperature show that little exchange interaction exist between the localized spins on the anions.

The correlation of the donor π-electron system and anion d-electron system is clearly evidenced in the magnetoresistance. Figure 4b presents the field dependence of the magnetoresistance of $(DMET)_2FeBr_4$ measured at 1.6 K ($<T_N$). The in-plane magnetoresistance $\rho_{//}$ takes a broad minimum around 6–7 T where the magnetization of the d-electron system is saturated. Moreover, when the external magnetic field is applied along the easy-axis direction, a discontinuous change appears at B_{SF} for the easy axis direction (// a axis). The interplane hopping frequency $\omega_{\perp}$ calculated from the interplane resistance shows a steep increase at B_{SF} for B // a and a kink at B_2 for B // b data. These one-to-one correspondences between the magnetoresistance and magnetization directly evidence the presence of the π–d interaction in this system.

This correlation in the anomalies between the magnetization and the magnetoresistance is observed not only for $(DMET)_2FeBr_4$ but also for $(EDTDM)_2FeBr_4$ with remarkable pressure effects. By the application of pressure this sulfur-analogue salt behaves as a metal down to 1.8 K above $p_C \sim 9.2$ kbar, and under the pressure around p_C the resistivity shows a remarkable anomaly around 4 K corresponding to

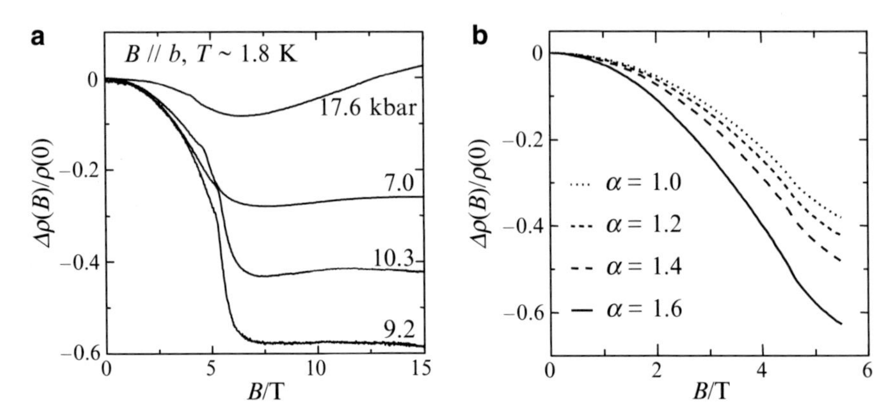

Fig. 5 (**a**) Field dependence of the intralayer magnetoresistance of $(EDTDM)_2FeBr_4$ at $T \sim 1.8$ K under pressures of 5.4, 7.0, 10.3, and 17.6 kbar. (**b**) Calculated field dependence of the magnetoresistance under various pressures, expressed with an empirical parameter α (see text)

the Néel temperature of the magnetic d-electron layers. The magnetoresistance in this salt measured at $T = 1.8$ K has large negative values (Fig. 5a) and becomes constant above ca. 6 T, which corresponds to the saturation field of the d-electron system. It should be noted that this negative magnetoresistance effect is the most significant around p_C, which suggests the importance of the quasi one-dimensional electronic instability in the appearance of this negative magnetoresistance effect.

These experimental results can be understood qualitatively as follows. Due to the quasi one-dimensional donor alignment, the π-electron system is the most sensitive to a perturbation by the wave vector $\boldsymbol{Q} \sim (\boldsymbol{a}^* \pm \boldsymbol{b}^*)/2$, which is the same as the wave vector of the antiferromagnetic spin alignment in the anion layer. Therefore the spin-alignment of the d-electron layers, which can be regulated by external magnetic field and/or temperature, strongly influences the transport properties inside the π-electron layer. From the theoretical point of view, this system is formulated by using the tight-binding Hamiltonian including the on-site Coulomb repulsion U and the π–d exchange interaction $J_{\pi d}$. Using this model Hamiltonian magnetoresistance under pressure (Fig. 5a) are semiquantitavely reproduced as shown in Fig. 5b, where the magnitude of $J_{\pi d}$ is estimated to be -0.3 K, and the pressure effect is introduced to the transfer integrals between chains using an empirical factor α as $t_{\perp}(p) = \alpha t_{\perp}(0)$. Therefore we can conclude that the quasi one-dimensional electronic instability is responsible for the large anomalous negative magnetoresistance of this material, which is amplified by the d-electron spin alignment through the π–d interaction.

Finally, we discuss the effects of chalcogen (S to Se) substitution. Although in $(EDS\text{-}TTF)_2FeBr_4$ remarkably close Se···Br contacts exist between donor and anion layers, little magnetic exchange interaction exists between the magnetic anions. The molecular orbital calculation reveals that the contribution of Se $4p$ orbital to the HOMO of the EDS-TTF molecule is considerably smaller than the contribution of

S $3p$ orbital to the HOMO of DMET or EDTDM molecule. Therefore $|t_{\pi d}|$ and also $|J_{\pi d}|$ of the EDS-TTF salt becomes negligibly small compared to the DMET or EDTDM salt. It should also be noted that the intermolecular Br···Br distance between $FeBr_4$ anions is almost the same among the EDS-TTF, DMET and EDTDM salts. Therefore this also proves the importance of the π–d interaction in the long-range magnetic ordering of the d-electron spins.

The two salts $(DMET)_2FeBr_4$ and $(EDTDM)_2FeBr_4$ show similar physical properties, but for $(DMET)_2FeBr_4$ system the pressure-induced phenomena such as negative magnetoresistance are less prominent than $(EDTDM)_2FeBr_4$. In DMET molecules, two of the four sulfur atoms in the central TTF unit are replaced with selenium atoms. This increases the intermolecular transfer integral t and reduces the on-site Coulomb repulsion U; hence the SDW state of $(DMET)_2FeBr_4$, which is responsible for the pressure-induced phenomena as discussed above, is less stabilized compared to $(EDTDM)_2FeBr_4$.

4 Effect of Halogen–Halogen Interactions in Molecular Conducting Magnets with Halogen-Substituted Donors

Our next approach to increase the magnitude of the π-d interaction is the introduction of intermolecular halogen-halogen contacts. The attractive interaction between the halogen substituents mainly results from an electrostatic Coulomb interaction and partial charge-transfer. However, the geometrical preference of the contacts [55] suggests the anisotropic interaction contributed from the lone-pair electron density of the halogen atoms. Namely, the donation of the lone-pair electrons from one halogen atom to the others gives the anisotropy of the interaction. Due to this covalent characteristic, the halogen-halogen contacts are evidently stronger than van der Waals contacts and thus are occasionally referred to as "halogen bonds" [56, 57] on the analogy of hydrogen bonds. As the halogen bonds are anisotropic, they have mainly been adopted in molecular conductors mainly from the viewpoint of controlling crystal structures, i.e., so-called crystal engineering [58–62]. We can anticipate that these halogen-halogen contacts having a weak covalent character can mediate exchange interaction between the conduction π-electrons on the donors and localized d-electrons on the anions. Semiempirical calculations [63] show that if the π–d interaction relies only on intermolecular van der Waals contacts, the magnitude of the interaction is at most of the order of maximum 1 K. On the other hand, if halogen-halogen contacts are introduced into the molecular conducting magnets, stronger π–d interaction through these semicovalent bonds can be expected [64–67].

We first selected the bromine-substituted TTF-type donor $EDT\text{-}TTFBr_2$ (4,5-dibromo-4′,5′-ethylenedithiotetrathiafulvalene) [68] to achieve strong π–d interaction [69]. This donor molecule has two bromine atoms that cause attractive interaction with a halogen atom of MX_4^- anions. In this donor molecule the opposite side of TTF moiety is fused with a sulfur-containing six-membered ring that expands the π-orbital and stabilizes metallic state. Then we focused on the oxygen-substituted analogue EDO-$TTFBr_2$ (4,5-dibromo-4′,5′-ethylenedioxo-tetrathiafulvalene) [70] as a donor molecule [71–73]. This molecule or its iodine-analogue are promising, since the coexistence of metallic conduction and ferromagnetic interaction is observed in (EDO-$TTFI_2)_2[M(mnt)_2]$ (M = Ni, Pt; mnt = maleonitriledithiolate) [74] and an anomalous metallic state having both itinerant and localized character of π-electron system was found in (EDO-$TTFBr_2)_3I_3$ [75].

The crystal structures of (EDT-$TTFBr_2)_2MX_4$ and (EDO-$TTFBr_2)_2MX_4$ are quite similar, although the space group symmetry is different in these two systems. However, this difference comes only from the conformation of terminal six-membered rings of the donor molecules, which plays no important role in the physical properties of the present salts. The donor molecules are stacked in a head-to-tail manner to form quasi-one-dimensional columns as shown in Fig. 6a.

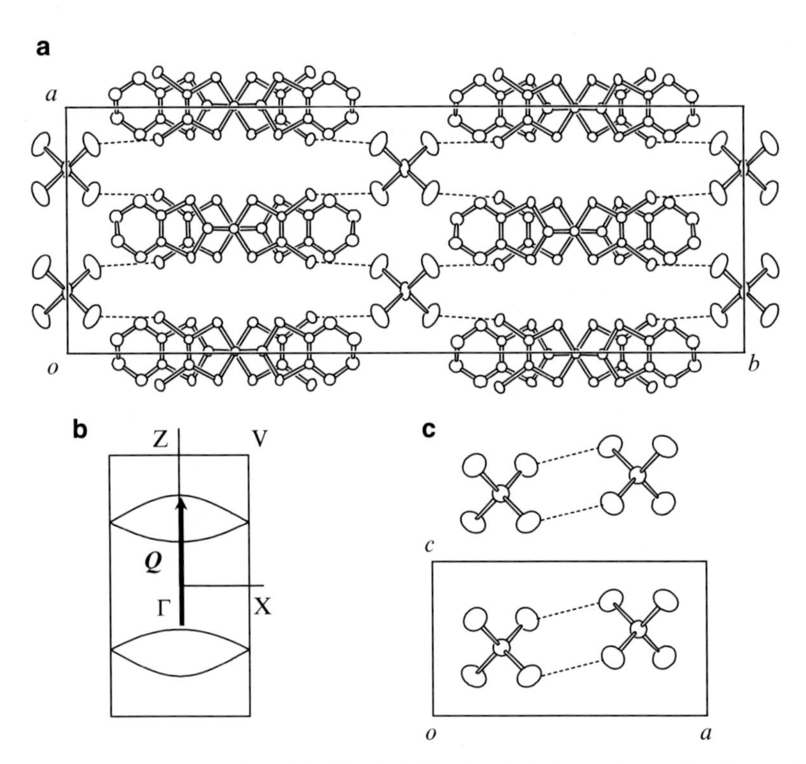

Fig. 6 (**a**) Crystal structure of (EDO-$TTFBr_2)_2FeCl_4$. The *dashed lines* denote close intermolecular donor-anion contacts. (**b**) Fermi surfaces obtained for a donor layer around $z = 1/2$ using the tight-binding approximation. The *solid arrow* represents the nesting vector $\boldsymbol{Q} \sim (\mathbf{c}^*)/2$. (**c**) Structure of the anion layer of (EDO-$TTFBr_2)_2FeCl_4$. The *dashed lines* are intermolecular anion-anion contacts

As these molecules lie on the special position determined by space group symmetry, the donor columns are uniform with no dimerization. The transfer integrals inside the columns are three to five times larger than the transfer integrals between the chains, showing that the donor layers are characterized as a quasi one-dimensional system. As a result, the Fermi surfaces obtained by the tight-binding band calculation have a lentil-like shape; thus the electronic nature of the present system can be characterized as a quasi one-dimensional metallic system with a possible nesting vector $\boldsymbol{Q} \sim \boldsymbol{c}^*/2$ as shown in Fig. 6b. The MX_4^- counter anions are sandwiched between the donor columns (Fig. 6c), and the halogen-halogen contacts along the side-by-side direction of the donor molecules have a distance slightly longer than the van der Waals distances. Therefore the direct anion-anion exchange interaction between the Fe^{3+} spins in the one-dimensional anion chains is very weak or almost absent. The most remarkable intermolecular contacts are Br···Br or Br···Cl contacts observed between the bromide group of the donors and the anion ligands, the distance of which is remarkably smaller than the corresponding van der Waals distance, especially in the case of (EDO-TTFBr$_2$)$_2$FeCl$_4$. For the EDT-TTFBr$_2$ salts, the intermolecular short S···Br contacts also strengthen the interaction between donor and anion molecules. Therefore in both EDT-TTFBr$_2$ and EDO-TTFBr$_2$ salts the long anion-anion distances and the short donor-anion distances suggest the importance of the π–d interaction in the physical properties of the salts.

The magnetic susceptibilities of (EDT-TTFBr$_2$)$_2$FeBr$_4$ and (EDO-TTFBr$_2$)$_2$ FeX$_4$ salts obey the Curie–Weiss law, and their negative Weiss temperatures indicate the presence of the antiferromagnetic exchange interaction between the d-electron spins. The χ-T plots of (EDT-TTFBr$_2$)$_2$FeBr$_4$ and (EDO-TTFBr$_2$)$_2$FeCl$_4$ show a broad peak around 20 and 7 K, respectively, due to the short-range order of d-electron spin (Fig. 7a, b), and then they undergo antiferromagnetic phase transitions at T_N = 11 and 4.2 K, respectively. Here we should remind that no significant halogen-halogen contacts are observed between the magnetic ions in these salts. Therefore the magnetic ordering in these salts must be realized with close anion-donor-anion contacts though the π–d interaction. In (EDT-TTFBr$_2$)$_2$FeBr$_4$ the Br···S contacts between the donor and anion layers also strengthen this π–d interaction. Consequently, the strong π–d interaction evidences the usefulness of the chemical modification of the donor molecule in these salts.

Compared to these two antiferromagnetic salts, the magnetic properties of the FeBr$_4$ salt are complicated (Fig. 7c). An antiferromagnetic transition firstly occurs at T_N = 13.5 K, and the susceptibility along the a-axis takes the maximum at T_{C2} = 8.5 K then begins to decrease as the temperature decreases. In the lowest temperature region, the susceptibilities along the a- and b-axes decrease and approach finite positive values, suggesting the presence of a helical-ordered spin state. This possibility is also supported by a model calculation in which the exchange interaction between the anions through the donor layer is taken into account. In other words, the complex magnetic behavior observed in this salt also results in the π–d interaction through the intermolecular halogen-halogen contacts.

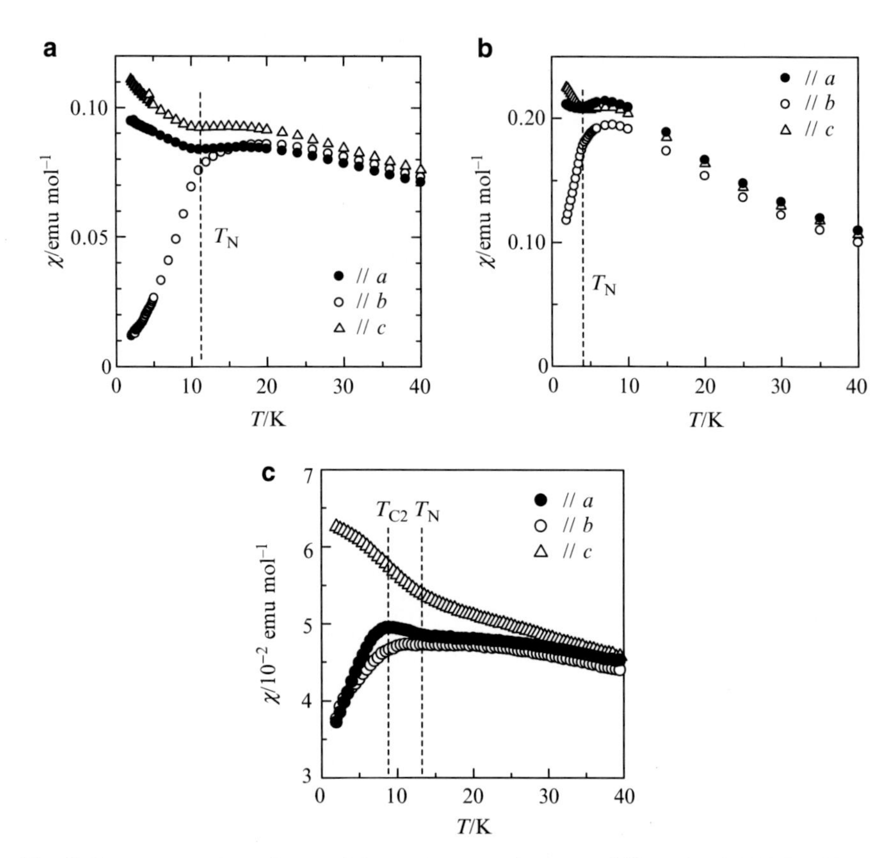

Fig. 7 Temperature dependence of the static magnetic susceptibility of: (**a**) $(EDT\text{-}TTFBr_2)_2FeBr_4$; (**b**) $(EDO\text{-}TTFBr_2)_2FeCl_4$; (**c**) $(EDO\text{-}TTFBr_2)_2FeBr_4$ measured at an external field of B = 1 T after the core diamagnetic contributions are subtracted

All of these three salts exhibit metallic behavior of the electrical conductivity around room temperature. In the low-temperature region, their metallic behavior gradually changes to semiconductive. From the ESR spectra and static magnetic susceptibilities of $GaBr_4$ or $GaCl_4$ salts having diamagnetic anions, the ground state of the π-electron layer is characterized as a Mott insulator for $(EDT\text{-}TTFBr_2)_2FeBr_4$, and SDW states for the two $EDO\text{-}TTFBr_2$ salts. The difference in the ground state can be understood from the intermolecular transfer integrals t. Due to the smaller van der Waals radius of oxygen than sulfur atom, the TTF unit of the $EDO\text{-}TTFBr_2$ salts are closely packed and thus have larger t than $EDT\text{-}TTFBr_2$ salts. Therefore the $EDT\text{-}TTFBr_2$ salts have larger U/t ratio which favors the Mott insulator states, whereas the $EDO\text{-}TTFBr_2$ salts have smaller U/t ratio which gives itinerant characters to the π-electrons. Nevertheless due to the quasi one-dimensional character of the donor layer, the ground state of the $EDO\text{-}TTFBr_2$ salts becomes SDW states. The metal-insulator transition is most remarkably observed in

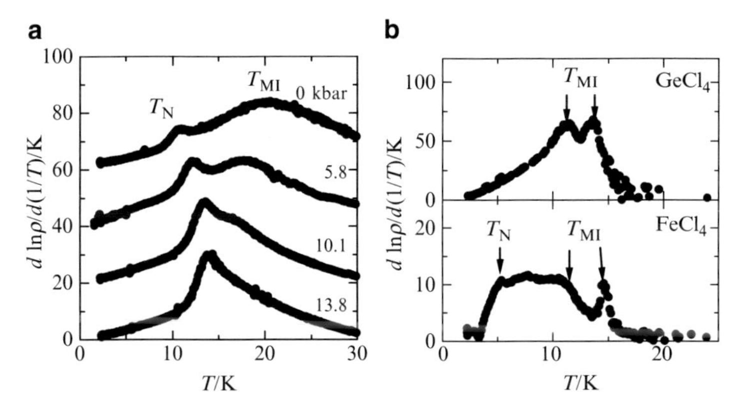

Fig. 8 Temperature dependence of $d\ln \rho/d(T^{-1})$, i.e., slope of the Arrhenius plot as a function of temperature for: (**a**) (EDT-TTFBr$_2$)FeBr$_4$ at various pressures – the data for 0, 5.8 and 10.1 kbar are vertically shifted up by 60, 40 and 20 K, respectively, for clarity; (**b**) (EDO-TTFBr$_2$)$_2$GaCl$_4$ and (EDO-TTFBr$_2$)$_2$FeCl$_4$ at 11 kbar. T_{MI} and T_N are the metal-insulator transition temperature and the Néel temperature, respectively. In (**b**) the metal–insulator transition is observed as two separate peaks

the plot of the activation energy $E_A = d\ln \rho/d(T^{-1})$ (ρ: resistivity) as a function of temperature as shown in Fig. 8. For (EDT-TTFBr$_2$)$_2$FeBr$_4$ the peak at T_{MI} corresponds to the metal-insulator transition, and the peak at T_N shows the magnetic transition of the counter anion layer (Fig. 8a). As the external pressure increases, T_{MI} decreases monotonically corresponding to the decrease in the U/t ratio that stabilizes the metallic state, whereas T_N increases monotonically and finally merges to T_{MI} at 13.8 kbar, resulting from the increase in the exchange interaction between the magnetic anions. For the EDO-TTFBr$_2$ salts (Fig. 8b), on the other hand, the metal-insulator transition is observed as doubly-split peaks, and the magnetic transition of the counter anion in (EDO-TTFBr$_2$)FeCl$_4$ is observed as a kink at the temperature T_N. These results indicate the interaction between the π- and d-electron systems.

Figure 9 shows the magnetoresistance of (EDT-TTFBr$_2$)$_2$FeBr$_4$ and (EDO-TTFBr$_2$)$_2$FeCl$_4$ under various pressures in their antiferromagnetic phase, applying the magnetic field along the spin-easy axis. The magnetoresistance of (EDT-TTFBr$_2$)$_2$FeBr$_4$ takes negative values with a kink, of which magnetic field coincides with the spin-flop field observed at the magnetization curves. As the pressure increases, this spin-flop transition field slightly increases then disappears above 10 kbar, although the anomaly of $d \ln \rho/d(T^{-1})$ vs T plot becomes more prominent in this pressure range (Fig. 8a). Possible explanation is the ineffectiveness of the d-electron spins in the π-electron transport in the high-pressure range where the metallic features are enhanced, which is supported by the reduction of the negative magnetoresistance as the pressure increases.

The magnetoresistance of (EDO-TTFBr$_2$)FeCl$_4$, on the other hand, shows multistep behaviors, as shown in Fig. 9b. In a low-field region, two stepwise

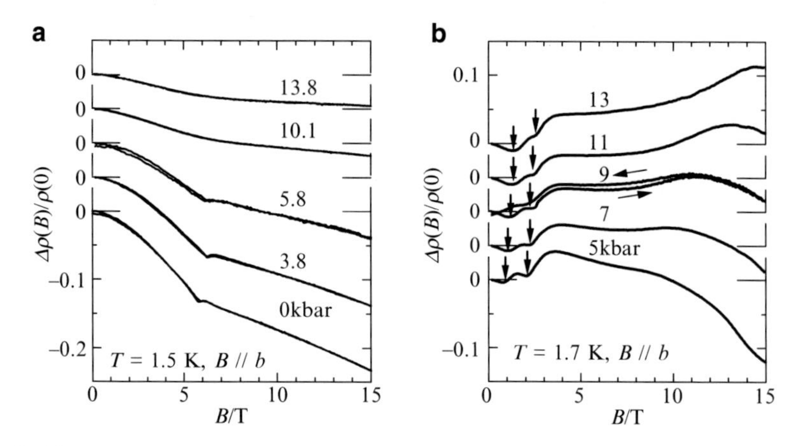

Fig. 9 Magnetoresistance of: (**a**) $(EDT\text{-}TTFBr_2)_2FeBr_4$; **b** $(EDO\text{-}TTFBr_2)_2FeCl_4$ under various pressures. In (**b**) the stepwise anomalies are indicated by i, and the hysteresis is shown for $p = 9$ kbar

increases are observed at $B_{SF-\pi} \sim 1$ T and $B_{SF-d} \sim 2$ T as indicated by arrows, which are attributed to the spin-flop transitions of the π-electron spins and d-electron spins, respectively. In the field above the saturation field $B_{Sat-d} \sim 12$T of the $FeCl_4$ anions, the magnetoresistance begins to decrease as the field increases. It should also be noted that a small hysteretic behavior of the magnetoresistance curve is observed in the field range of 0–10 T (in the figure, only the data at 9 kbar are shown for simplicity). This complex behavior in the magnetoresistance can be qualitatively explained as the frustration of the triangles composed of one EDO-$TTFBr_2$ molecule and two neighboring $FeCl_4$ anions that affects the SDW state of the donor layer. This frustration diminishes successively after the spin flop of the donor and anion layers at $B_{SF-\pi}$ and B_{SF-d}, respectively, which stabilizes the SDW state to increase the resistivity. When the field reaches the saturation field of the d-electron spins (B_{Sat-d}), all the anion spins align along the magnetic field direction and the stabilization of the SDW state through the π–d interaction is lost; therefore, the magnetoresistance decreases above this field.

5 Conclusion and Prospect

We have developed molecular conducting magnets based on TTF-type donors and magnetic counter anions by adopting two strategies to increase the effectiveness of the π–d interaction with respect to the physical properties. The ground state of the quasi one-dimensional π-electron system in $(DMET)_2FeBr_4$ and $(EDTDM)_2FeBr_4$ are characterized as the SDW state. For the latter salt it changes from the SDW state to the metallic state by applying pressure, and near the pressure-induced insulator-metal transition point, the spin alignment of the magnetic anion layer affects the transport properties of the organic donor layer. The crystal structures of $(EDT\text{-}TTFBr_2)_2FeBr_4$ and $(EDO\text{-}TTFBr_2)_2FeX_4$ (X = Br, Cl) are characterized by the presence of close intermolecular halogen-halogen contacts between the donor and anion layers, which function as exchange interaction paths to produce the magnetic ordered state of the anion spins and magnetotransport properties in the donor π-electron system affected by the anion spins.

Although we have succeeded in strengthening the π–d interaction by means of halogen-halogen contacts, its magnitude (ex. 22 K in $(EDT\text{-}TTFBr_2)_2FeBr_4$ [69]) is still too small compared to the exchange interaction between π-electrons inside the conduction layers (ex. 1,400 K in $(TMTSF)_2PF_6$ [16]). For the further enhancement of the π–d interaction it will be useful to introduce chemical bonds, such as coordination or covalent bonds, between the π- and d-electron components. This "through-bond" approach has been investigated using the direct coordination of paramagnetic metal ions to organic radicals having pyridine [76–82], phosphine [83–86] and acetylacetone [87] substituents. Another approach is the introduction of metal-carbon covalent bond between these components. Metal-carbon bonds have strong bonding character, in contrast to the antibonding character of coordination bonds, and thus expected to give stronger π–d interaction. For example, the d-electron spin in paramagnetic organoiron complexes $[(\eta^2\text{-dppe})(\eta^5\text{-}C_5Me_5)Fe^{III}\text{–}C{\equiv}C\text{–}Ar]^+$ (dppe: 1,2-bis(diphenylphosphino)ethane, Ar: aryl group) is significantly delocalized into the aryl group [88], and thus binuclear compounds having these organometallic units connected with carbon-rich spacers show strong magnetic exchange interaction between Fe spins separated by more than 10 Å [89]. Focusing on this remarkable characteristic, we are now working on TTF derivatives connected with this organometallic unit to produce novel π–d interaction-based molecular conducting magnets. It is expected that the molecular conducting magnets developed by these novel strategies including through-bond approaches will give new viewpoints in the development of molecular electronics and/or spintronics devices.

References

1. Kagoshima S, Kanoda K, Mori T (eds) (2006) Organic conductors, special topics section. J Phys Soc Jpn 75
2. Batail P (ed) (2004) Molecular conductors, special issue. Chem Rev 104
3. Ishiguro T, Yamaji K, Saito G (1998) Organic superconductors. Springer, Berlin Heidelberg New York
4. Akamatu H, Inokuchi H, Matsunaga Y (1956) Bull Chem Soc Jpn 29:213–218
5. Uchida T, Akamatu H (1961) Bull Chem Soc Jpn 34:1015–1020
6. Gaim AK, Novoselov KS (2007) Nat Mater 6:183–191
7. Ferrais J, Cowan DO, Walatka V Jr, Perlstein JH (1973) J Am Chem Soc 95:948–949
8. Denoyer F, Comès R, Garito A, Heeger A (1975) Phys Rev Lett 35:445–449
9. Jérome D, Mazaud A, Ribault M, Bechgaard K (1980) J Phys Lett 41:L95–L98

10. Bechgaard K, Carneiro K, Olsen M, Rasmussen FB, Jacobsen CS (1981) Phys Rev Lett 46:852–855
11. Grüner G (1994) Density waves in solids, frontiers in physics. Addison-Wesley, Reading, MA, USA
12. Kino H, Fukuyama H (1996) J Phys Soc Jpn 65:2158–2169
13. Takahashi T, Nogami Y, Yakushi K (2006) J Phys Soc Jpn 75:051008
14. Tomonaga S (1950) Prog Theor Phys 5:544–569
15. Luttinger JM (1963) J Math Phys 4:1154–1162
16. Dumm M, Loidl A, Fravel BW, Starkey KP, Montgomery LK, Dressel M (2000) Phys Rev B 61:511–521
17. Lee IJ, Brown SE, Naughton MJ (2006) J Phys Soc Jpn 75:051011
18. Sawano F, Terasaki I, Mori H, Mori T, Watanabe M, Ikeda N, Nogami Y, Noda Y (2005) Nature 437:522–524
19. Yamada J, Sugimoto T (2004) TTF chemistry: fundamentals and applications of tetrathiafulvalene. Springer, Berlin Heidelberg New York
20. Taniguchi H, Miyashita M, Uchiyama K, Satoh K, Mori N, Okamoto H, Miyagawa K, Kanoda K, Hedo M, Uwatoko Y (2003) J Phys Soc Jpn 72:468–471
21. Miyazaki A, Yamazaki H, Aimatsu M, Enoki T, Watanabe R, Ogura E, Kuwatani Y, Iyoda M (2007) Inorg Chem 46:3353–3366
22. Kudo S, Miyazaki A, Enoki T, Golhen S, Ouahab L, Toita T, Yamada J (2006) Inorg Chem 45:3718–3725
23. Okabe K, Yamaura JI, Miyazaki A, Enoki T (2005) J Phys Soc Jpn 74:1508–1520
24. Setifi F, Ouahab L, Golhen S, Miyazaki A, Okabe K, Enoki T, Toita T, Yamada J (2002) Inorg Chem 41:3786–3790
25. Miyazaki A, Okabe K, Enomoto K, Nishijo J, Enoki T, Setifi F, Golhen S, Ouahab L, Toita T, Yamada J (2003) Polyhedron 22:2227–2234
26. Enomoto K, Yamaura J, Miyazaki A, Enoki T (2003) Bull Chem Soc Jpn 76:945–953
27. Nishijo J, Ogura E, Yamaura J, Miyazaki A, Enoki T, Takano T, Kuwatani Y, Iyoda M (2003) Synth Met 133–134:539–541
28. Enoki T, Yamazaki H, Okabe K, Nishijo J, Enomoto K, Enomoto M, Miyazaki A (2002) Mol Cryst Liq Cryst 379:131–140
29. Miyazaki A, Enomoto M, Enomoto K, Nishijo J, Enoki T, Ogura E, Kuwatani Y, Iyoda M (2002) Mol Cryst Liq Cryst 376:535–542
30. Enomoto M, Miyazaki A, Enoki T (2001) Bull Chem Soc Jpn 74:459–470
31. Nishijo J, Miyazaki A, Enoki T, Ogura E, Takano T, Kuwatani Y, Iyoda M, Yamaura J (2000) Solid State Commun 116:661–664
32. Miyazaki A, Umeyama T, Enoki T, Ogura E, Kuwatani Y, Iyoda M, Nishikawa H, Ikemoto I, Kikuchi K (1999) Mol Cryst Liq Cryst 334:379–388
33. Enoki T, Umeyama T, Enomoto M, Yamaura J, Yamaguchi K, Miyazaki A, Ogura E, Kuwatani Y, Iyoda M, Kikuchi K (1999) Synth Met 103:2275–2278
34. Miyazaki A, Enomoto M, Enomoto M, Enoki T, Saito G (1997) Mol Cryst Liq Cryst 305:425–434
35. Yamaura J, Suzuki K, Kaizu Y, Enoki T, Murata K, Saito G (1996) J Phys Soc Jpn 65:2645–2654
36. Enoki T, Miyazaki A (2004) Chem Rev 104:5449–5477
37. Ouahab L, Enoki T (2004) Eur J Inorg Chem 2004:933–941
38. Miyazaki A, Enomoto K, Okabe K, Yamazaki H, Nishijo J, Enoki T, Ogura E, Ugawa K, Kuwatani Y, Iyoda M (2002) J Solid State Chem 168:547–562
39. Enoki T, Yamaura J, Miyazaki A (1997) Bull Chem Soc Jpn 70:2005–2023
40. Coronado E, Day P (2004) Chem Rev 104:5419–5448
41. Wang M, Xiao X, Fujiwara H, Sugimoto T, Noguchi S, Ishida T, Mori T, Aruga-Katori H (2007) Inorg Chem 46:3049–3056
42. Naito T, Inabe T (2004) Bull Chem Soc Jpn 77:1987–1995

43. Hanasaki N, Matsuda M, Tajima T, Ohmichi E, Osada T, Naito T, Inabe T (2006) J Phys Soc Jpn 75:033703
44. Coronado E, Galán-Mascarós JR, Gómez-García CJ, Laukhin V (2000) Nature 408:447–449
45. Kobayashi H, Cui HB, Kobayashi A (2004) Chem Rev 104:5265–5288
46. Kobayashi H, Kobayashi A, Cassoux P (2000) Chem Soc Rev 29:325–333
47. Mohn P (2003) Magnetism in the solid state. Springer, Berlin Heidelberg New York
48. Ruderman MA, Kittel C (1954) Phys Rev 96:99–102
49. Kasuya T (1956) Prog Theor Phys 16:45–57
50. Yosida K (1957) Phys Rev 106:893–898
51. Deaton JC, Gebhard MS, Solomon EI (1989) Inorg Chem 28:877–889
52. Shannon RD (1969) Acta Crystallogr A 32:751–767
53. Miyazaki A, Enoki T (2003) Synth Met 133/134:543–545
54. Bondi A (1964) J Phys Chem 68:441–451
55. Desiraju GR, Parthasarathy R (1989) J Am Chem Soc 111:8725–8726
56. Legon C (1999) Angew Chem Int Ed Engl 38:2687–2714
57. Ouvrard C, Le Questel JY, Berthelot M, Laurence C (2003) Acta Crystallogr B 59:512–526
58. Domercq B, Devic T, Fourmigué M, Auban-Senzier P, Canadell E (2001) J Mater Chem 11:1570–1575
59. Imakubo T, Shirahata T, Hervé K, Ouahab L (2006) J Mater Chem 16:162–173
60. Suizu R, Imakubo T (2003) Org Biomol Chem 1:3629–3631
61. Imakubo T, Tajima N, Tamura M, Kato R, Nishio Y, Kajita K (2002) J Mater Chem 12:159–161
62. Imakubo T, Maruyama T, Sawa H, Kobayashi K (1995) Chem Commun 1667–1668
63. Mori T, Katsuhara M (2002) J Phys Soc Jpn 71:826–844
64. Hervé K, Cador O, Golhen S, Costuas K, Halet JF, Shirahata T, Muto T, Imakubo T, Miyazaki A, Ouahab L (2006) Chem Mater 18:790–797
65. Ouahab L, Setifi F, Golhen S, Imakubo T, Lescouëzec R, Lloret F, Julve M, Świetlik R (2005) C R Chim 8:1286–1297
66. Thoyon D, Okabe K, Imakubo T, Golhen S, Miyazaki A, Enoki T, Ouahab L (2002) Mol Cryst Liq Cryst 376:25–32
67. Imakubo T, Sawa H, Kato R (1996) Mol Cryst Liq Cryst 285:27–32
68. Kux U, Suzuki H, Sasaki S, Iyoda M (1995) Chem Lett 14:183–184
69. Nishijo J, Miyazaki A, Enoki T, Watanabe R, Kuwatani Y, Iyoda M (2005) Inorg Chem 44:2493–2506
70. Iyoda M, Kuwatani Y, Ogura E, Hara K, Suzuki H, Takano T, Takeda K, Takano J, Ugawa K, Yoshida M, Matsuyama H, Nishikawa H, Ikemoto I, Kato T, Yoneyama N, Nishijo J, Miyazaki A, Enoki T (2001) Heterocycles 54:833–848
71. Enoki T, Yamazaki H, Nishijo J, Ugawa K, Ogura E, Kuwatani Y, Iyoda M, Sushko YV (2003) Synth Met 137:1173–1174
72. Miyazaki A, Aimatsu M, Yamazaki H, Enoki T, Ugawa K, Ogura E, Kuwatani Y, Iyoda M (2004) J Phys IV 114:545–547
73. Miyazaki A, Aimatsu M, Enoki T, Watanabe R, Ogura E. Kuwatani Y, Iyoda M (2006) J Low Temp Phys 142:477–480
74. Nishijo J, Miyazaki A, Enoki T, Ogura E, Takano T, Kuwatani Y, Iyoda M, Yamaura JI (2000) Solid State Commun 116:661–664
75. Miyazaki A, Kato T, Yamazaki H, Enoki T, Ogura E, Kuwatani Y, Iyoda M, Yamaura JI (2003) Phys Rev B68:085108
76. Gavrilenko KS, Le Gal Y, Cador O, Golhen S, Ouahab L (2007) Chem Commun 2007:280–282
77. Benbellat N, Gavrilenko KS, Le Gal Y, Cador O, Golhen S, Gouasmia A, Fabre JM, Ouahab L (2006) Inorg Chem 45:10440–10442
78. Hervé K, Liu SX, Cador O, Golhen S, Le Gal Y, Bousseksou A, Stoeckli-Evans H, Decurtins S, Ouahab L (2006) Eur J Inorg Chem 2006:3498–3502

79. Ota A, Ouahab L, Golhen S, Cador O, Yoshida Y, Saito G (2005) New J Chem 29:1135–1140
80. Liu SX, Ambrus C, Dolder S, Neels A, Decurtins S (2006) Inorg Chem 45:9622–9644
81. Liu SX, Dolder S, Franz P, Neels A, Stoeckli-Evans H, Decurtins S (2003) Inorg Chem 42:4801–4803
82. Devic T, Rondeau D, Sahin Y, Levillain E, Clérac R, Batail P, Avarvari N (2006) Dalton Trans 2006:1331–1337
83. Uzelmeier CE, Smucker BW, Reinheimer EW, Shatruk M, O'Neal AW, Fourmigué M, Dunbar KR (2006) Dalton Trans 2006:5259–5268
84. Gouverd C, Biaso F, Cataldo L, Berclaz T, Geoffroy M, Levillain E, Avarvari N, Fourmigué M, Sauvage FX, Wartelle C (2005) Phys Chem Chem Phys 7:85–93
85. Avarvari N, Fourmigué M (2004) Chem Commun 2004:1300–1301
86. Avarvari N, Martin D, Fourmigué M (2002) J Organomet Chem 643/644:292–300
87. Massue J, Bellec N, Chopin S, Levillain E, Roisnel T, Clérac R, Lorcy D (2005) Inorg Chem 44:8740–8748
88. Paul F, da Costa G, Bondon A, Gauthier N, Sinbandhit S, Toupet L, Costuas K, Halet JF, Lapinte C (2007) Organometallics 26:874–896
89. Lapinte C (2008) J Organomet Chem 693:793–801

Top Organomet Chem (2009) 27: 97–140

Metallocenium Salts of Transition Metal Bisdichalcogenate Anions; Structure and Magnetic Properties

V. Gama and M. Almeida

Abstract The properties of the salts based on metallocenium cations and transition metal bisdichalcogenide anions are reviewed and particular attention is paid to the correlation between the magnetic properties correlated and the crystal and molecular structures. The large majority of these salts have crystal structures based on linear chain arrangements of alternating cation (D^+) and anion (A^-) stacks which are classified in four major structural types, depending on the stacking motifs. The magnetic properties of these salts, which at low temperatures can present a wide range of magnetic order type and phase transitions, are correlated with the type of magnetic interaction between the magnetic building blocks, which in the large majority of the cases can be well described by the McConnell model using spin density calculations.

Keywords Crystal structures, Magnetic properties, Metallocenium salts, Metal-bisdithiolenes

Contents

V. Gama and M. Almeida(✉)
Department Química, Instituto Tecnológico e Nuclear/CFMCUL, P-2686-953 Sacavém, Portugal,
E-mail: malmeida@itn.pt

M. Fourmigué and L. Ouahab (eds.), *Conducting and Magnetic Organometallic Molecular Materials*, Topics in Organometallic Chemistry 27,
DOI: 10.1007/978-3-642-00408-7_5,

Abbreviations

A^-	Anion
AF	Antiferromagnet (ic, ism)
bds	Benzene-1, 2-diselenolate
bdt	Benzene-1, 2-dithiolate
Cp	Cyclopentadienyl
Cp*	Pentanethylcyclopentadienyl
D^+	Cation
dcdmp	2, 3-Dicyano-5, 6-dimercaptopyrazine
dmio	1, 3-Dithiol-2-one-4, 5-dithiolate
dmit	1, 3-Dithiol-2-thione-4, 5-dithiolate
dsit	2-Thioxo-1, 3-dithiole-4, '5-diselenolate
edt	Ethylenedithiolate
EPR	Electron paramagnetic resonance
FAP	Field aligned paramagnet
FIM	Ferrimagnet (ic, ism)
FM	Ferromagnet (ic, ism)
MM	Metamagnet (ic, ism)
mnt	Maleonitriledithiolate
SCE	Saturated Calomel electrode
SF	Spin-flop
TCNE	Tetracyanoethylene
TCNQ	7, 7, 8, 8-Tetracyano-*p*-quinodimethane
tcbdt	3, 4, 5, 6-Tetracholorobenzene-1, 2-dithiolate
tds	(Trifluoromethyl)ethylene diselenolate
tdt	(Trifluoromethyl)ethylene dithiolate
α-tpdt	2, 3-Thiophenedithiolate

1 Introduction

Both metallocenes and transition metal bis(1, 2-dithiolene) complexes are important building blocks that have been increasingly used during the last few years for the preparation of molecule based magnetic materials. The importance of these building blocks for the design of molecular materials derives from their great versatility as units that, depending on the metal and substituents, can bear different magnetic moments and can have tunable physical properties. Indeed, the possibility of preparing families of structurally and electronically related compounds based on these units has allowed systematic relationships between structure and magnetic properties of molecular materials to be deduced. In this chapter the materials

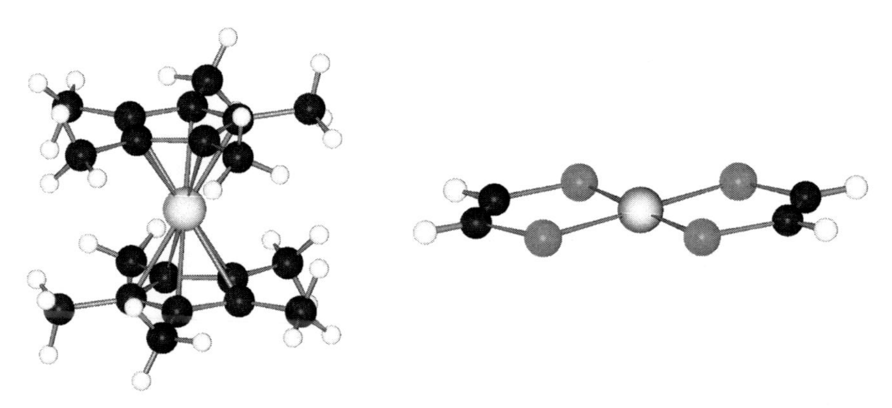

Fig. 1 Molecular structure of $[M(Cp^*)_2]^+$ and $[Ni(edx)_2]^-$ (edx = 1, 2-ethylenedichalcogenate), the basic cation and anions in this review

obtained as charge transfer salts by the combination of these two type of molecular units, the first as metallocenium cations and the last as anions, will be reviewed, with emphasis on the correlation between their crystal structure and magnetic properties (Fig. 1).

The metallocenes with the general formula $M(C_5R_5)_2$ and consisting of two cyclopentadieneyl anions (Cp) bound to a metal in the center can exist in different oxidation states, where the redox potentials depend on the metal M and on the groups R. One of the most commonly used cyclopentadienyl is the pentamethyl substituted, $Cp(CH_3)_5$ or Cp*, that stabilizes the metallocenes allowing different oxidation states. The mono cationic $[M(Cp^*)_2]^+$ states are thus obtained for a variety of metals M, namely from groups 6–9 with different stability. Depending on the metal M, these $[M(Cp^*)_2]^+$ cations present different magnetic moments, atomic spin density distribution and magnetic ion anisotropy. The spin states S are due to the electrons distribution in the metal d-orbitals, which are split by D_{5h} or D_{5d} ligand field symmetry, for the staggered and eclipsed configuration respectively. Most frequently used are third row transition metals M = Co ($S = 0$), Fe, Ni ($S = 1/2$), Mn ($S = 1$), Cr, V ($S = 3/2$), which often lead to isostructural compounds and where the effect of the magnetic moment magnitude can easily be tested. However, the stability of the corresponding cations is different and the oxidation potentials of the couple $[M(Cp^*)_2]/[M(Cp^*)_2]^+$ increases along the series Fe, Mn, Ni, Cr, o (−0.12, −0.56, −0.65, −1.04, −1.47 V vs. SCE respectively) [1, 2] in such a way that, with the exception of Fe, the manipulation of all other metallocenium requires the employment of strict anaerobic conditions, in glove box or by Schlenk techniques, to their manipulation. In addition to the different spin values these cations can present different ion anisotropies as denoted by the g-values of the EPR signal. For the S=1/2 $[Fe(Cp^*)_2]^+$ cation a large anisotropy is found with $g_{\parallel} = 4.4$ and $g_{\perp} = 1.25$, $\langle g \rangle = 2.8$ [3], for $[Mn(Cp^*)_2]^+$ a large anisotropy is also expected in view of the dispersion of g values in the range 2.2–2.9 [2, 4–8] obtained from the magnetic measurements with polycrystalline samples, but the S=3/2 $[Cr(Cp^*)_2]^+$ presents virtually no magnetic anisotropy with $g_{\parallel} = g_{\perp} = 2.0$ [1].

The potential of metallocenium ions in molecular magnetic materials was recognized early and dates at least to the discovery of the first molecule-based material exhibiting ferromagnetic ordering, the salt $[Fe(Cp^*)_2]$TCNE (TCNE = tetracyanoethylene), with $T_C = 4.8$ K, in 1986 [9, 10]. Since this landmark discovery in molecular magnetism, other metallocenium based salts have received considerable attention [11–13]. Soon after the discovery of ferromagnetism in $[Fe(Cp^*)_2]$ TCNE the transition metal bisdichalcogenate planar anions were also considered as suitable candidates for preparing new molecular magnets, and the first salts of metallocenium cations and transition metal bisdichalcogenate anions were reported in 1989 [14, 15].

The transition metal-bis(1, 2-dithiolene) complexes have been extensively used for more than 30 years as building blocks for the preparation of both conducting and magnetic molecular materials [16, 17]. Some of the most commonly used dithiolenes ligands are depicted in Scheme 1. Due to the active contribution of the ligands to the frontier orbitals, depending on the ligands these complexes can exist in a variety of oxidations states which can range from $z = -2$ to 0, and in same cases can even go to cationic states. Partial oxidation situations are not uncommon among transition metal-bisdithiolene complexes, particularly with the solid state being the basis of unique electrically conducting properties.

These complexes exist in a diversity of coordination geometries, the square planar being preferred by metals of groups 10 and 11. Other coordination

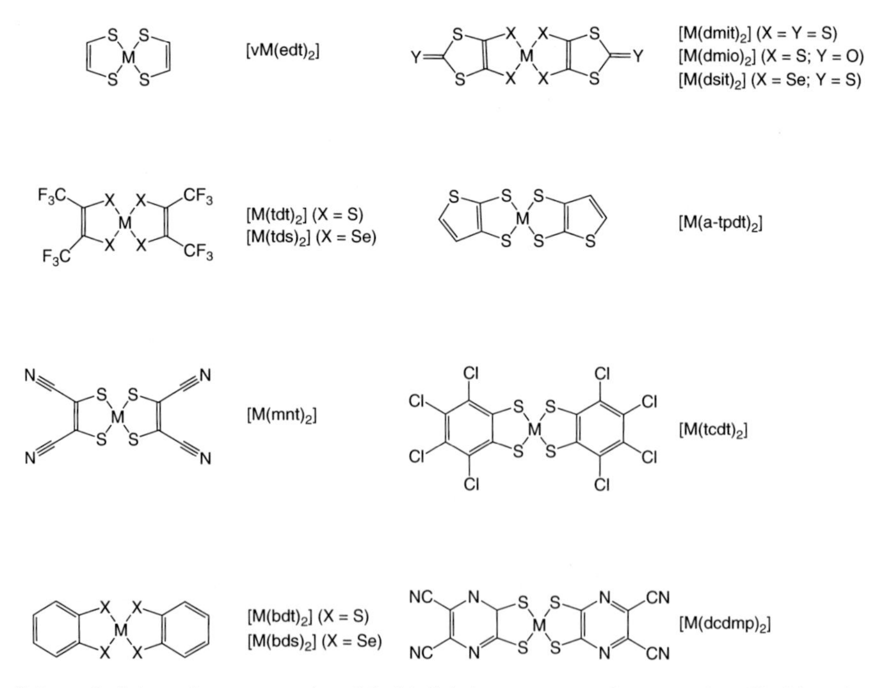

Scheme 1 Schematic representation of the bisdichalcogenate complexes mentioned in this review

geometries are favored by different metals, such as for instance Fe and Co, which have a preference for dimeric arrangements of the M-bis(dithiolene) units with the metal with a square-pyramidal coordination geometry [17–19]. For Co^{III}, further to this dimeric structure, examples of trimeric [20] or even polymeric structures are also known [21]. The use of extended π-ligands and square planar geometries has been actively explored for the application of these complexes in solid state materials because the planar configuration favors the stacking of molecules in the solid state maximizing intermolecular interactions through either π−π or S. . .S interactions. The key feature of these complexes in the design of magnetic materials is that, depending on the oxidation state and on transition metal, different spin states are possible. Most used have been the monoanionic complexes of group 10 metals (Ni, Pd and Pt) which are $S = 1/2$, and of group 11 metals (Cu and Au) which are diamagnetic. Monoanionic complexes of groups 8 and 9 metals (Fe and Co), which are usually dimeric, can present high or low spin states depending on the ligand.

Several transition metal bisdithiolene complexes are associated with significant contributions for the development of molecular magnetism. When combined with the perylene radical cations in the $Per_2[M(mnt)_2]$ compounds, the $[M(mnt)_2]^-$ chains (mnt = 1, 2-dicyano-1, 2-ethylene-dithiolato), with M = Ni, Pd, Pt, provided the first example of a multifunctional conducting and magnetic material where localized spin chains provided by stacks of $M(mnt)_2$ anions coexisted with chains of delocalized conducting electrons, in a rather unique situation in solid state materials [22, 23]. The competing spin-Peierls and Peierls instabilities with these two types of chains has been a topic of active research and recently these materials protagonized the discovery of a series of field induced states [24, 25]. Bisdithiolene complexes based on more xtended and sulfur rich ligands, such as dmit, were at the basis of the first inorganic based molecular superconductors [26, 27].

The nature of the magnetic interactions between paramagnetic transition metal bisdithiolene complexes is quite sensitive to variations in the geometry of molecular overlap and contacts. This is specially important in larger ligands due to spin polarization effects extending negative or positive spin densities throughout the more extended ligands. Most of the simple bisdithiolene salts with magnetic innocent cations present antiferromagnetic interactions in solid state. However, some salts such as n-$Bu_4N[Ni(\alpha\text{-tpdt})_2]$ were found to present ferromagnetic interactions [28]. The simple dithiolene salt $NH_4[Ni(mnt)_2]\cdot H_2O$ was shown to display ferromagnetic ordering at low temperatures and under hydrostatic pressure [29, 30] and more recently also $BrFBzPy[Ni(mnt)_2]$ at ambient pressure [31]. More recent work reported bisdithiolene complexes based on extended ligands incorporating TTF moieties that in their neutral state are simultaneously magnetic and conducting [32, 33].

This review concerns the materials made from the combination as charge transfer salts of the above-mentioned metallocenium cations with transition metal bisdithiolene anions, expanding and updating a previous review [34]. Most of the materials studied so far are decamethylmetallocenium based salts, but other compounds based on different metallocenium derivatives have also been reported and will be also referred.

As magnetic ordering is a bulk property, a particular attention will be given to the supramolecular arrangements in these compounds, which determine the magnetic behavior and to the correlation between structure and magnetic properties. The magnetic coupling between units in this type of salts has been analyzed mainly through McConnell I [35] or McConnell II [36] mechanisms, and this issue is still a subject of controversy [37, 38]. Between those models, McConnell I has been the most used in the interpretation of the magnetic behavior of these salts, as, in spite of its simplicity, it has shown a good agreement with the experimental observations. Therefore in this work the analysis of the magnetic coupling will often be performed under that model. However, it should be mentioned that the validity of McConnell I mechanism has been questioned both theoretically [39] as well as experimentally [40].

2 Basic Structural Motives

2.1 *Decamethylmetallocenium-Based Salts*

Most salts based on decamethylmetallocenium cations and metal bisdichalcogenene anions, due to the planar configuration of the both the C_5Me_5 ligands and of the anions, present crystal structures based on linear chain arrangements of alternating cation (D^+) and anion (A^-) stacks. The structures based on these alternated chains can be classified essentially in four basic distinct types as schematically depicted in Fig. 2. Type I corresponds to the most simple case of an alternated linear chain motive $\cdot\cdot A^-D^+A^-D^+A^-D^+\cdot\cdot$, similar to that observed in several salts based on metallocenium cations and on acceptors such as TCNE and TCNQ [11–13]. In type II chains, the cations alternate with face-to-face pairs of anions, $\cdot\cdot A^-A^-D^+A^-A^-D^+\cdot\cdot$. As in this arrangement there is a net charge (−) per repeat unit, $A^-A^-D^+$, charge compensation by an additional cation is required. Type III consists of alternated face-to-face pairs of anions with side-by-side pairs of cations, $\cdot\cdot A^-A^-D^+D^+A^-A^-D^+D^+\cdot\cdot$. Finally, in type IV arrangement, the anions alternate with side-by-side pairs of cations, $\cdot\cdot A^-D^+D^+A^-D^+D^+\cdot\cdot$, and in this case the net charge (+) per repeat unit, $D^+D^+A^-$, requires an additional anion for compensation. In this type IV structural motive, three variants based on different orientation of the pairs of cation were detected, as illustrated in Fig. 2b. In compounds based on types II and IV arrangements (with charged chains), the different possibilities concerning the position of compensation ions lead to a diversified range of structural variations, while with salts based on types I and III arrangements (neutral chains), basically only one type of chain arrangement is observed. Table 1 summarizes the spin of the cations and anions, the basic structural motives, the Weiss constants, ϑ, the critical temperatures, and magnetically ordered states of salts based on decamethylmetallocenium and on metal bisdichalcogenate anions.

While in the case of the cyano radical based salts, most of the observed structures present a type I structural arrangement, in the case of the metal bisdichalcogenate-based

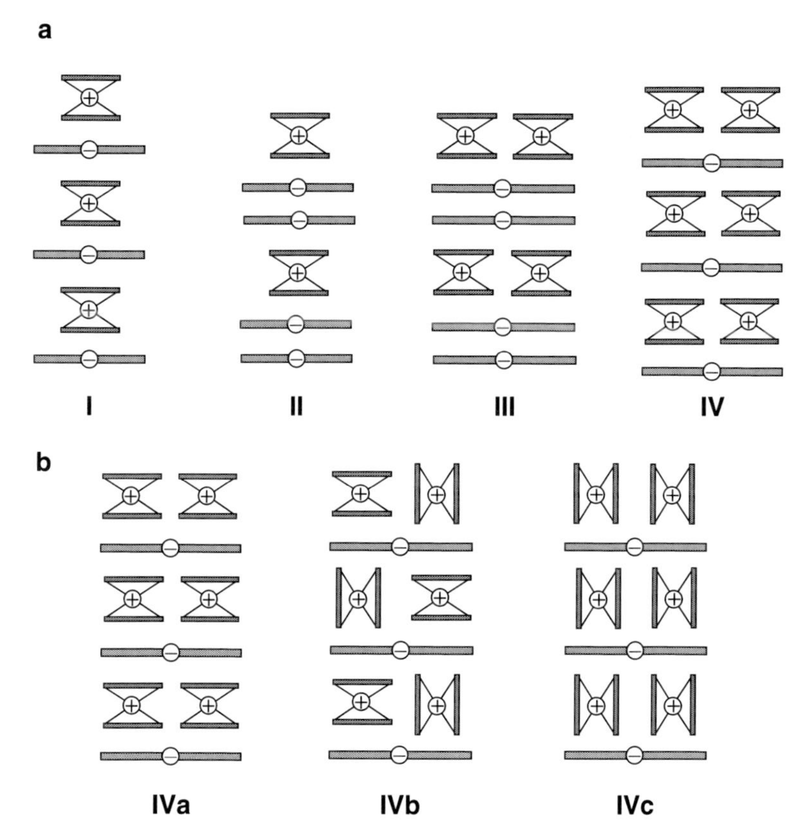

Fig. 2 **(a)** Basic structural mixed chain motifs of the metallocenium based salts of metal bisdichalcogenate anions. **(b)** Variations of the type IV chain motif

salts, a much larger variety of arrangements was observed, as described above. The structural motives in the $[M(Cp^*)_2][M'(L)_2]$ salts are primarily determined by factors such as the dimensions of the anionic metal bisdichalcogenate complexes, the tendency of the anions to associate as dimers, the degree of extension of the π system in the anions, and the charge density distribution on the ligands.

In case of $[M(edt)_2]^-$ based salts, the size of the small anion is similar to the size of the C_5Me_5 ligand of the cation and only type I structural motives ($D^+ A^- D^+ A^-$ chains) were observed. For the intermediate size anionic complexes, $[M(tdx)_2]^-$, $[M(mnt)_2]$ and $[Ni(\alpha\text{-tpdt})_2]^-$, the most common structural motive obtained in salts based on those anions is also of type I. For the larger anionic complexes, $[M(bdx)_2]^-$ and $[M(dmix)_2]^-$, types III and IV chain arrangements were observed. In both cases anion molecules (type IV) or face-to-face pair of anions (type III) alternate with side-by-side pairs of cations. The complexes $[M(mnt)_2]^-$ and $[M(dmix)_2]^-$ (M = Ni, Pd and Pt) frequently present dimerization in the solid state [19], and they are the only anions where the chain arrangements present face-to-face pairs of anions (structural motives II and III). The variety of structural

Table 1 Cation and anion S values; basic structural motives; Weiss constants; transition temperatures, magnetic ordering; critical fields of the salts based on metallocenium and on metal bisdichalcogenate complexes

Compound (chain type)	θ, K	T_C, K	Magnetic behavior[a]	H_C,kG[b] (T, K)	Ref
$[Fe(Cp^*)_2][Ni(edt)_2]$ (I)	−5	4.2	MM	14 (2)	[41]
$[Cr(Cp^*)_2][Ni(edt)_2]$ (I)	−6.7				[41]
$[Fe(Cp^*)_2][Ni(tdt)_2]$ (I)	15				[14]
$[Mn(Cp^*)_2][Ni(tdt)_2]$ (I)	2.6	2.4	MM		[42]
$[Mn(Cp^*)_2][Pd(tdt)_2]$ (I)	3.7	2.8	MM	0.8 (1.85)	[42]
$[Fe(Cp^*)_2][Pt(tdt)_2]$ (I)	27				[11–13, 43]
$[Mn(Cp^*)_2][Pt(tdt)_2]$ (I)	1.9	2.3	MM	–	[42]
$[Fe(Cp^*)_2][Ni(tds)_2]$ (I)	8.9				[44–46]
$[Mn(Cp^*)_2][Ni(tds)_2]$ (I)	12.8	2.1	MM	0.28 (1.6)	[45, 46]
$[Cr(Cp^*)_2][Ni(tds)_2]$ (I)	2.3				[45, 46]
$[Fe(Cp^*)_2][Pt(tds)_2]$ (I)	9.3	3.3	MM	3.95 (1.7)	[44–46]
$[Mn(Cp^*)_2][Pt(tds)_2]$ (I)	16.6	5.7	MM	4.05 (1.7)	[45, 46]
$[Cr(Cp^*)_2][Pt(tds)_2]$ (I)	9.8	5.2	AF/SF[c]	5; 16 (1.7)	[45, 46]
$[Fe(Cp^*)_2][Ni(\alpha\text{-tpdt})_2]$ (I)	3.8	2.6	MM	0.6 (1.6)	[28, 47]
$[Mn(Cp^*)_2][Ni(\alpha\text{-tpdt})_2]$ (I)	7.5	~4	Frust. Mgn.	−0.5 (1.7)[d]	[47]
$[Cr(Cp^*)_2][Ni(\alpha\text{-tpdt})_2]$ (I)	3				[47]
$[Co(Cp^*)_2][Ni(\alpha\text{-tpdt})_2]$ (I)	−4.2				[48]
$[Fe(Cp^*)_2][Au(\alpha\text{-tpdt})_2]$ (I)[e]	−6.2				[48]
$[Fe(C_5Me_4SCMe_3)_2][Ni(mnt)_2]$ (I)	3				[49]
α-$[Fe(Cp^*)_2][Pt(mnt)_2]$ (I)	6.6				[14]
β-$[Fe(Cp^*)_2][Pt(mnt)_2]$ (I)	9.8				[14]
$[Fe(C_5Me_4SCMe_3)_2][Pt(mnt)_2]$ (I)	3				[49]
$[Fe(Cp^*)_2]_2[Cu(mnt)_2]$ (IVc)	−8.0				[50]
$[Fe(Cp^*)_2][Ni(dmit)_2]$ (III)	−7.6[f]				[15, 51]
$[Mn(Cp^*)_2][Ni(dmit)_2]$ (III)	2.5	2.5	FIM	3.5 G (2)[g]	[52, 53]
[Co(Cp*)2][Ni(dmit)2] (III)	0.5				[54]
α-$[Fe(Cp^*)_2][Pd(dmit)_2]$ (III)	−22.3[h]				[55, 56]
β-$[Fe(Cp^*)_2][Pd(dmit)_2]$ (III)[e]	2.6[f]				[51, 55, 56]
$[Fe(Cp^*)_2][Pt(dmit)_2]$ (III)	−14.4[h]				[51, 55, 56]
$[Mn(Cp^*)_2][Au(dmit)_2]$ (III)	−4.2				[52]

$[Fe(Cp^*)_2][Ni(dmio)_2]$ (III)[e]	-19.0[h]				[51]
$[Fe(Cp^*)_2][Ni(dmio)_2]MeCN$ (IVa)	2.0				[57]
$[Fe(Cp^*)_2][Ni(dmio)_2]\cdot THF$ (IVa)	10.5				[58, 59]
$[Mn(Cp^*)_2][Ni(dmio)_2]Me_2CO$ (IVa)	8.5				[54]
[Mn(Cp*)2][Ni(dmio)2]MeCN (IVa)	2.8				[52]
$[Mn(Cp^*)_2]_2[Ni(dmio)_2]_2PhCN$ (III)	-3.62				[54]
$[Fe(Cp^*)_2][Pd(dmio)_2]$ (III)	-24.7[h]				[51]
$[Fe(Cp^*)_2][Pt(dmio)_2]$ (III)	-33.3[h]				[51]
$[Fe(Cp^*)_2][Ni(dsit)_2]$ (III)	-18.9[h]				[60]
$[Fe(Cp^*)_2][Ni(bdt)_2]$ (IVb)	-5.6[f]				[55, 61, 62]
$[Mn(Cp^*)_2][Ni(bdt)_2]$ (IVb)	-22.6[f]	2.3	MM/FIM[i]	0.2 (2)	[55, 61, 62]
$[Cr(Cp^*)_2][Ni(bdt)_2]$ (IVb)	+6.2				[55, 61, 62]
$[Fe(Cp^*)_2][Co(bdt)_2]$ (IVb)[e]	-21.6[f]				[55, 61]
$[Mn(Cp^*)_2][Co(bdt)_2]$ (IVb)	-8.8[f]				[55, 61]
$[Cr(Cp^*)_2][Co(bdt)_2]$ (IVb)	-7.7[f]				[55, 61]
$[Fe(Cp^*)_2][Pt(bdt)_2]$ (IVb)	-12.6[f]				[55, 61]
$[Mn(Cp^*)_2][Pt(bdt)_2]$ (IVb)	-20.5[f]	2.7	FIM		[55, 61]
$[Cr(Cp^*)_2][Pt(bdt)_2]$ (IVb)	+6.0				[55, 61]
$[Fe(Cp^*)_2][Ni(bds)_2]MeCN$ (IVa)	0[f]				[15]
$[Fe(Cp^*)_2]_2[Cu(dcdmp)_2]$ (IVa)	~0				[63]
$[Fe(Cp^*)_2][Ni(tcbdt)_2]$[j]	-22.9				[62]
$[Mn(Cp^*)_2][Ni(tcbdt)_2]$[j]	-28.5				[62]
$[Cr(Cp^*)_2][Ni(tcbdt)_2]$[j]	-20.4				[62]

[a] MM (metamagnet), AF (antiferromagnet), SF (spin-flop), Frust. Mgn. (frustrated magnet) and FIM (ferrimagnet)
[b] Critical field, unless sated otherwise
[c] Class II AF, with a SF field induced transition from the AF ground state
[d] Reversed coercivity in the isothermal hysteresis loop
[e] Expected structural motif, crystal structure not determined
[f] No CW behavior, θ value obtained at high temperatures
[g] Coercive field
[h] No CW behavior, θ value obtained with poor fits
[i] FIM high field state
[j] Crystal structure not determined

arrangements observed in the $[M(mnt)_2]^-$ based compounds can be related both with the large extension of the π system and with the high charge density on the terminal nitrile groups [64], as well as to the tendency of these complexes to form dimers.

2.2 *Salts Based on other Metallocenium Cations*

Besides the decamethylmetallocenium salts, in the compounds based on other metallocenium derivatives, mixed linear chain arrangements were only observed in the case of the salts $[Fe(C_5Me_4SCMe_3)_2][M(mnt)_2]$, M = Ni and Pt, which present type I structural motives.

Some salts based on other metallocenium derivatives and on the anions $[M(mnt)_2]^-$ and $[M(dmit)_2]^-$ (M = Ni and Pt) were also reported. In the case of these compounds, the crystal structure consists of segregated stacks of cations, $\cdot\cdot D^+D^+D^+D^+\cdot\cdot$, and anions, $\cdot\cdot A^-A^-A^-A^-\cdot\cdot$, which is a common situation in molecular materials, in particular in the case of molecular conductors. In spite of the fact that for most salts the dominant magnetic interactions between the metal bisdichalcogenate units are antiferromagnetic, there are cases where those interactions are known to be ferromagnetic, as in the case of the compounds n-$Bu_4N[Ni(\alpha\text{-tpdt})_2]$ [28] and $NH_4[Ni(mnt)_2]\cdot H_2O$, which was the first metal bisdichalcogenate-based material to present ferromagnetic ordering, with $T_C = 4.5$ K [29, 30]. The anion and cation S values and the dominant magnetic interactions of the ferrocenium derivative salts with crystal structures based on segregated anion stacks are summarized in Table 2.

3 Solid-State Structures and Magnetic Behaviors

In this section the salts based on metallocenium cations and metal bisdichalcogenate anions will be reviewed according to the previously referred structural classification. After referring to the general characteristics of the crystal structures the supramolecular features will be correlated with the magnetic properties.

3.1 *Type I Mixed Chain Salts*

With a few exceptions, the crystal structure of the salts based on the type I mixed chains consists of parallel arrangements of those chains. In most cases the magnetic behavior of these salts is dominated by ferromagnetic (FM) interactions, due to intrachain D^+ A^- coupling. Several of these salts exhibit metamagnetic (MM) behaviors due to the coexistence of weaker antiferromagnetic (AF) intrachain interactions.

Table 2 Cation and anion *S* values; Weiss constants of the metallocenium salts of metal bisdichalcogenate complexes based on segregate chains of anions

Compound	S_D; S_A	θ, *K*	Ref
$[Fe(Cp)_2]_2[Ni(mnt)_2]_2[Fe(Cp)_2]$	1/2; 1/2	NR[a]	[65]
$[Fe(C_5Me_4SMe)_2][Ni(mnt)_2]$	1/2; 1/2	≤0	[49]
$[Fe(C_5H_4R)_2][Ni(mnt)_2]$[b]	1/2; 1/2	<0	[66]
$[Fe(Cp)(C_5H_4CH_2NMe_3)][Ni(mnt)_2]$	1/2; 1/2	<0	[67]
$[Fe(Cp)(C_5H_4CH_2NMe_3)][Pt(mnt)_2]$	1/2; 1/2	<0	[67]
$[Co(Cp)_2][Ni(dmit)_2]$	0; 1/2	NR[a]	[68]
$[Co(Cp)_2][Ni(dmit)_2]_3$2MeCN	0; 1/2	NR[a]	[69]
[Fe(Cp)-R1-Fe(Cp)][Ni(mnt)2][c]	1/2; 1/2	<0	[70]
[Fe(Cp)-R2-Fe(Cp)][Ni(mnt)2][d]	1/2; 1/2	<0	[70]
[*n*-butylferrocene][Ni(mnt)2]	1/2; 1/2	<0	[71]
[*tert*-butylferrocene][Ni(mnt)2]	1/2; 1/2	<0	[71]
[1,1′-diethylferrocene][Ni(mnt)2]	1/2; 1/2	<0	[71]
[1,1′-diisopropylferrocene][Ni(mnt)2]	1/2; 1/2	<0	[71]
$[Fe(Cp^*)_2][Ni(mnt)_2]$[e]	1/2; 1/2	~0	[14]

[a]NR = not reported
[b]$[Fe(C_5R)_2]^+$ = 1,1′-bis[2-(4-(methylthio)-(E)-ethenyl]ferrocenium
[c][Fe(Cp)-R1-Fe(Cp)] = [1′,1″-R2-1,1″-biferrocene], R = isopropyl
[d][Fe(Cp)-R1-Fe(Cp)] = [1′,1″-R2-1,1″-biferrocene], R = dineopentyl
[e]Structure not 1D

The analysis of the crystal structures, the magnetic behaviors and atomic spin density calculations of several salts based on decamethylmetallocenium and on metal-bis(dichalcogenate) anions with structures consisting of arrangements of parallel alternating D^+A^- liner chains, in particular for the salts $[M(Cp^*)_2[M'(L)_2]$ (with L = tds, tdt and edt), allowed a systematic study of the intra- and interchain magnetic interactions [44, 45]. In these compounds a spin polarization is observed in the metallocenium cations but not in the anions with these small ligands (tds, tdt and edt). The analysis of the intrachain contacts in the perspective of the McConnell I mechanism suggests the existence of intrachain FM coupling, through the contacts involving the metal or chalcogen atoms (positive spin density) from the anions and the C atoms (negative spin density) from the Cp ring of the cations, which shows a good agreement with the experimental observations. A variety of interionic interchain contacts were observed in these salts, A^-A^- (Se-Se, S-S and C-C), D^+ D^+ (Me-Me) and D^+A^- (Me-S), and all these contacts were observed to lead to AF interchain coupling. A strict application of the McConnell I model was not possible in the case of the interchain contacts, as the shortest contacts would involve mediation through H or F atoms, which are expected to present a very small spin density [44, 45]. However the results regarding the nature of the interchain magnetic coupling would be compatible with that model if the contacts involving H or F atoms were neglected, as all the atoms involved in these contacts present a positive spin density. This analysis revealed that metamagnetism, which was observed in several compounds presenting a crystal structure consisting on a parallel arrangement of alternated 1D chains, is expected to occur in the other compounds presenting a similar solid state structure, where the metal bisdichalcogenate anions does not present spin polarization effects.

3.1.1 $[M(Cp^*)_2]$ $[M'(tdx)_2]$

The structurally related salts $[M(Cp^*)_2][M'(tds)_2]$ (M = Fe, Mn, Cr; M′ = Ni, Pt) and $[Fe(Cp^*)_2][Pt(tds)_2]$ allowed a systematic study of the effect of a diversity of variables on the magnetic behavior of these compounds, such as the variation of the spin of the cation, the role of the single ion anisotropy, the effect of the variation of the size of atoms involved in the intermolecular contacts. Furthermore, the analysis of the intermolecular contacts in these compounds provided a reasonable interpretation of the intra and interchain magnetic coupling, and its relative strength within the series [44, 45].

The salts $[Fe(Cp^*)_2][Pt(tdt)_2]$ [43], $[M(Cp^*)_2][Ni(tds)_2]$, with M = Fe, Mn and Cr; and $[M(Cp^*)_2][Pt(tds)_2]$, with M = Fe, Mn [44, 45], are isostructural and their crystal structure consist of a parallel arrangement of type I chains. In these chains the Ni or Pt atoms sit above the Cp fragments from the cations and intrachain D^+A^- contacts (c1) shorter than or of the order of the sum of the van der Waals radii, d_W, were observed involving Ni or Pt and C atoms from the Cp rings of the cations. A view normal to the chains is shown in Fig. 3a for $[Fe(Cp^*)_2][Pt(tds)_2]$. For this series of compounds the shortest interchain interionic separation was found in the in-registry pair I-II, shown in Fig. 3b, and it corresponds to an AA Se-Se contact (c2), with a distance only slightly larger than d_W (~5–15%). In the other interchain

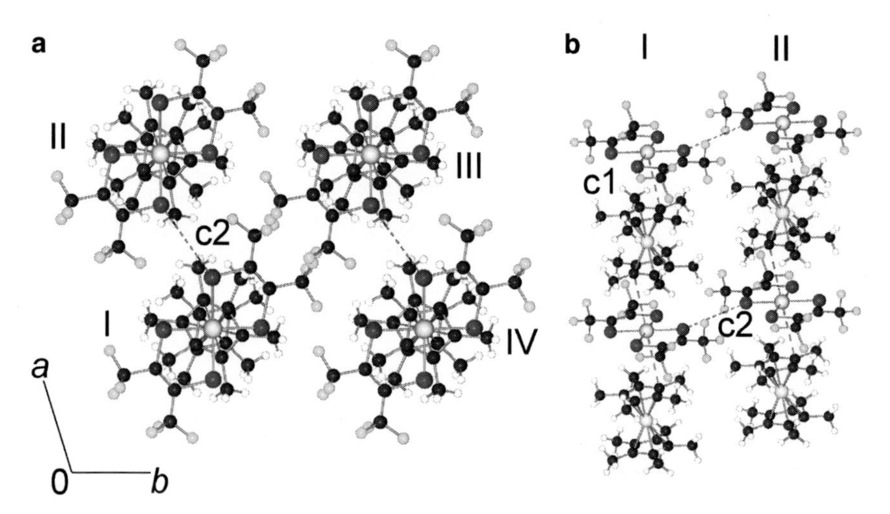

Fig. 3 (a) View of the crystal structure of $[Mn(Cp^*)_2][Pt(tds)_2]$ along the chain direction. **(b)** Interchain arrangement of the pair I–II, c1 corresponds to the closest intrachain contact (D^+A^-) and c2 to the closest interchain contact (A^-A^-)

arrangements the interchain contacts are considerably larger. In spite of $[Cr(Cp^*)_2]$ $[Pt(tds)_2]$ not being isostructural with the other compounds from the series, the intra and interchain arrangements are very similar to the above described for other salts, $[M(Cp^*)_2][M'(tds)_2]$ [45, 46].

The magnetic behavior of the compounds $[Fe(Cp^*)_2][Pt(tdt)_2]$, $[M(Cp^*)_2]$ $[Ni(tds)_2]$ and $[M(Cp^*)_2][Pt(tds)_2]$ (M = Fe, Mn and Cr) is clearly dominated by the strong intrachain D^+ A^- FM coupling, as can be seen by the positive θ values (Table 1). At low temperatures and low applied magnetic fields several of these compounds ($[Mn(Cp^*)_2][Ni(tds)_2]$, $[M(Cp^*)_2][Pt(tds)_2]$, M = Fe, Mn, Cr) show AF transitions and metamagnetic like behaviors upon the application of high enough magnetic fields. It was observed that the nature of these transitions depends on the magnetic anisotropy of the cations. The salts based on strongly anisotropic cations, such as $[M(Cp^*)_2]^+$, with M = Fe, Mn, present a phase diagram typical of Class I AF [72], while $[Cr(Cp^*)_2][Pt(tds)_2]$, where the $[Cr(Cp^*)_2]^+$ cation does not exhibit magnetic anisoptropy, shows a phase diagram typical of a Class II AF [72].

The magnetization temperature dependence at various values of the applied magnetic field is shown Fig. 4a for $[Mn(Cp^*)_2][Pt(tds)_2]$. At low applied magnetic fields a maximum occurs at around 5.5 K, indicating an AF phase transition. This maximum shifts to lower temperatures with increasing fields and is suppressed for high fields, due to a field induced transition. A T_N value of 5.7 K was obtained from a.c. susceptibility measurements. The magnetization isothermals, shown in Fig. 4b, present a clear sigmoidal behavior below T_N, which is typical of metamagnets. For $T \leq 3.2$ K the field induced transitions are quite sharp and began to exhibit hysteresis (increasing upon cooling), which can easily be seen in the 1.6 K isothermal, where both the measurements with increasing (closed squares; solid line) and decreasing (open squares; dashed line) applied fields are shown. This behavior indicates that below 3.2 K the field induced transition is a first-order phase transition, with $3.2 \leq T_T$ 4 K, where T_T is the tricritical temperature. An *H*,*T* phase diagram obtained from both d.c. magnetization, $M(T)$ and $M(H)$, and a.c. susceptibility, $\chi'(T)$ and $\chi'(H)$, results is shown in Fig. 5. This diagram includes two distinct magnetic phases, an AF and a paramagnetic phase, PM. Within the PM phase, at high fields and low temperatures an FM like region, "FM," is also defined and corresponds to the high field state induced by the applied magnetic field. Similar behavior is observed in the case of $[Mn(Cp^*)_2][Ni(tds)_2]$ and $[Fe(Cp^*)_2][Pt(tds)_2]$, exhibiting lower T_N values of 2.1 and 3.3 K respectively.

The temperature dependence of the magnetization of $[Cr(Cp^*)_2][Pt(tds)_2]$ is shown in Fig. 6a. At low applied magnetic fields an AF phase transition occurs, corresponding to a maximum at ~5 K. At low fields the magnetization decreases with cooling; however in the $M(T)$ curves with fields of 5 and 10 kG, after the maxima the magnetization passes through a minimum and increases slightly upon cooling. At 20 kG this maximum is no longer detected. Unlike the observed in the $[M(Cp^*)_2]^+$ (M = Fe and Mn) based compounds, the magnetization isothermal curves of $[Cr(Cp^*)_2][Pt(tds)_2]$, shown in Fig. 6b, reveal the existence of two field induced transitions, one occurring at around 4.9 kG (I) and a second one at higher applied magnetic fields (II), as indicated by the applied field dependence of d*M*/d*H*,

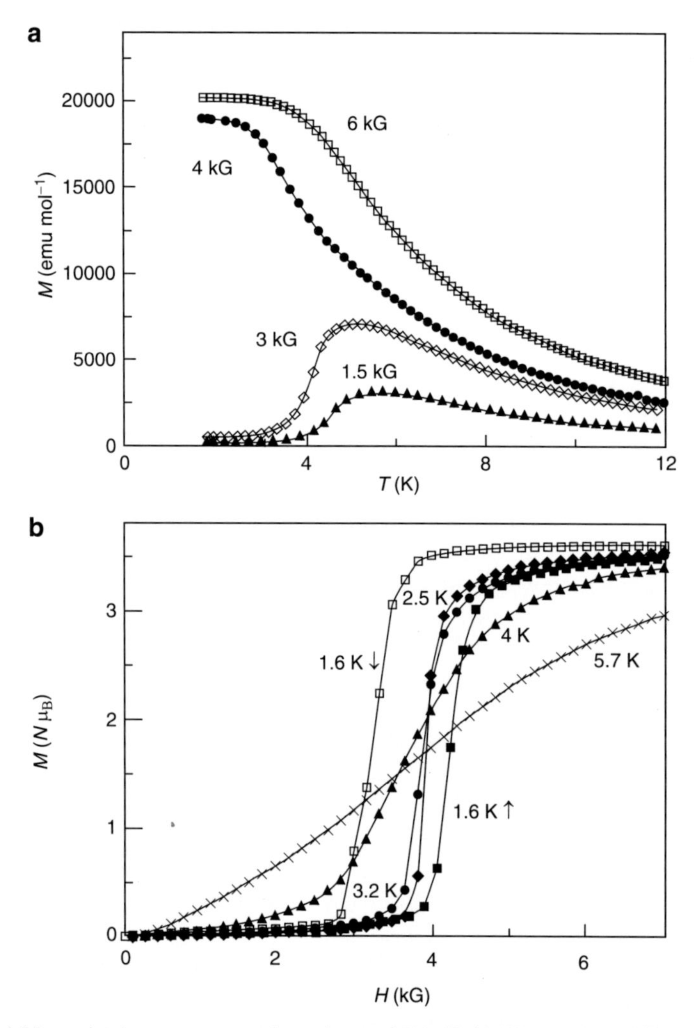

Fig. 4 **(a)** Magnetization temperature dependence of [Mn(Cp*)$_2$][Pt(tds)$_2$], at different magnetic fields. **(b)** Magnetization isothermals of [Mn(Cp*)$_2$][Pt(tds)$_2$], at different temperatures indicated. From [45]

shown in the inset of Fig. 6b. Two maxima in dM/dH, corresponding to the two field induced transitions, are observed. The peak in dM/dH corresponding to transition I presents a weak temperature dependence and at 4 K is no longer detected, while transition II shows a significant temperature dependence. At low temperatures, the d.c. magnetization behavior clearly suggest the existence of an AF phase at low fields ($H < 4.9$ kG) and a spin-flop (SF) phase at higher magnetic fields ($4.9 < H < 16$ kG). The (H,T) phase diagram of [Cr(Cp*)$_2$][Pt(tds)$_2$], obtained from both the d.c. magnetization, $M(T)$ and $M(H)$, and the real component of the a.c. susceptibility, $\chi'(T)$ and $\chi'(H)$ data, is shown in Fig. 7. This diagram shows three distinct

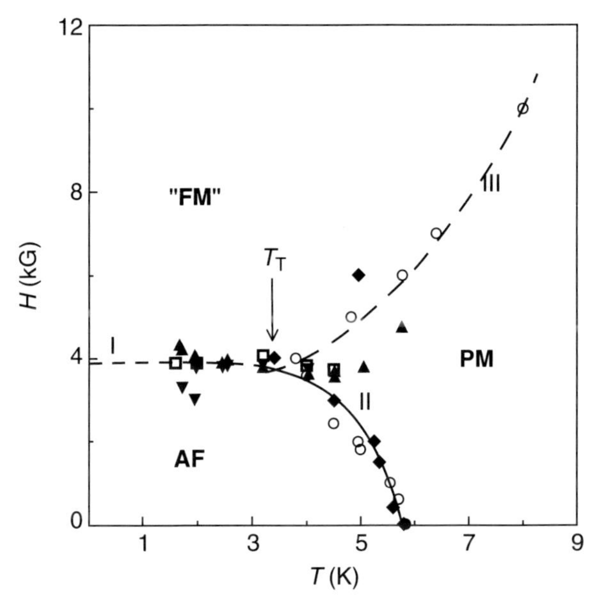

Fig. 5 Magnetic phase diagram of $[Mn(Cp^*)_2][Pt(tds)_2]$; $M(T)$ (*filled diamonds*) $M(H)$ ($H\uparrow$ (*filled triangles*), $H\downarrow$ (*filled inverted triangles*), $\chi'(T)$ (*open circles*) $\chi'(H)$ (*open squares*); T_T is the tricritical temperature; I denotes the first-order MM transition; II denotes a second-order transition (AF-PM phase boundary) and III denotes a higher order transitions (from a PM to a FM like state). From [45]

phases: an AF phase, an SF phase and a paramagnetic phase, PM. The SF (AF-SF) transition is first-order, while the AF-PM (III) and SF-PM transitions are expected to be second-order (phase boundaries represented by the solid line in Fig. 12). The first-order SF phase transition (phase boundary corresponds to the dashed line) meets the PM phase boundary at the triple point, T_t ($T \sim 3.1$ K; $H \sim 4.9$ kG). This compound is the first example of a decamethylmetallocenium salt exhibiting an SF phase transition.

The magnetic behavior of $[Fe(Cp^*)_2][Pt(tdt)_2]$ and $[M(Cp^*)_2][M'(tds)_2]$ (M = Fe, Mn, Cr; M′ = Ni, Pt) is consistent with the coexistence of FM intrachain interactions with weaker with weaker AF interchain interactions. The nature of the intra- and interchain magnetic interaction is in good agreement with the predictions of McConnell I mechanism [44–46]. The strong FM intrachain interactions can be assigned to short contacts (d1) between the Ni or Pt atoms from the anions (with a positive spin density, $\rho^S > 0$) and the C atoms from the Cp rings of the cations (with $\rho^S < 0$). While the AF interchain contacts are associated with the S. . .S or Se. . .Se contacts (d2) between anions in adjacent chains. The replacement of Se or Pt by the smaller S or Ni atoms leads to significant decrease in the interchain or in the intrachain coupling, as the chalcogen atoms are directly involved in the interchain contacts and the metals in the intrachain contacts. This effect is attributed to the weakening of the overlap in those contacts [44].

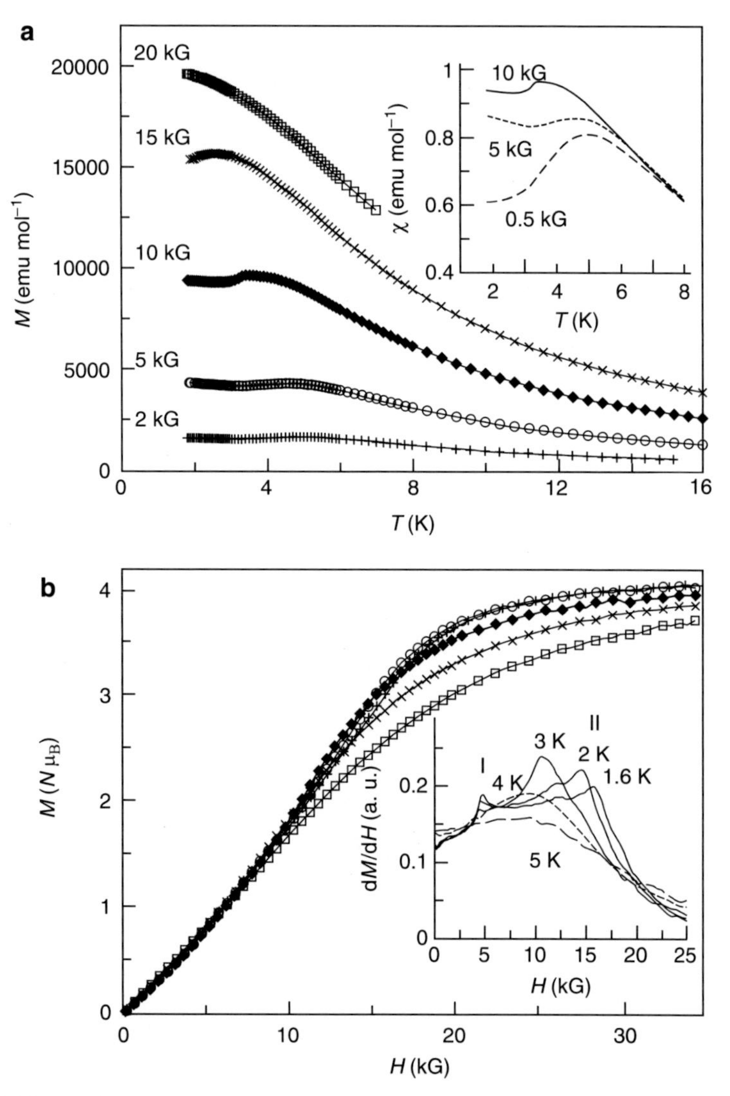

Fig. 6 **(a)** Magnetization temperature dependence of $[Cr(Cp^*)_2][Pt(tds)_2]$, under different magnetic fields. In the *inset*: χ vs T (with applied fields of 0.5, 5 and 10 kG). **(b)** Magnetization isothermals of $[Cr(Cp^*)_2][Pt(tds)_2]$, at different temperatures. In the *inset*: dM/dH vs H (at the same temperatures). From [45]

The salts $[Fe(Cp^*)_2][Ni(tdt)_2]$ and $[Mn(Cp^*)_2][M'(tdt)_2]$ (M′ = Ni, Pd and Pt) are isostructural, presenting a crystal structure based on an arrangement of parallel type I chains [14, 42]. These chains are not regular and the anions, nearly perpendicular to the chain axis, present two distinct orientations, with an angle of ~62°. The cations show a considerable tilting relative to the chain axis (~22°), the Cp rings sitting above the NiS_2C_2 rings of the anions, as shown for $[Fe(Cp^*)_2][Ni(tdt)_2]$ in Fig. 8.

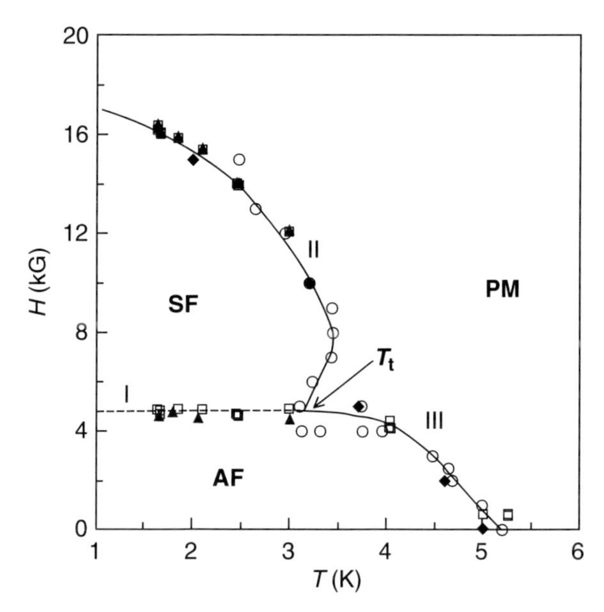

Fig. 7 Magnetic phase diagram of compound $[Cr(Cp^*)_2][Pt(tds)_2]$: *M*(*T*) (*filled diamonds*), *M*(*H*) (*filled triangles*), $\chi'(T)$ (*open circles*) $\chi'(H)$ (*open squares*); T_t is the triple point; I denotes the first-order SF transition; II and III denote second-order transitions (SF–PM and AF–PM phase boundaries). From [45]

In this compound, no intrachain D^+A^- short contacts were found and the closest interatomic separation between the anion and the Cp ring corresponds to Ni...C contacts (c1) exceeding the sum of the van der waals radii, d_W, by ~11%. In this salt the closest interchain arrangements are similar and present an out-of-registry disposition (pairs I–II and I–IV). This arrangement is shown in Fig. 8b. The closer interchain contacts (d2) involve C atoms from the Me groups of the cations and S atoms of the anions, with a separation exceeding d_W by ~8%.

The magnetic behavior of the compounds $[Fe(Cp^*)_2][Ni(tdt)_2]$ and $[Mn(Cp^*)_2]$ $[M'(tdt)_2]$, with M′ = Ni, Pd and Pt, is dominated by the intrachain D^+A^- FM interactions, as seen by the positive θ values (Table 1). At low temperatures the [Mn $(Cp^*)_2][M'(tdt)_2]$ salts exhibit metamagnetic transitions, with T_N = 2.4, 2.8 and 2.3 K for M′ = Ni, Pd and Pt respectively, H_C = 800 G at 1.85 K for M′ = Pd [42]. This behavior is attributed to the coexistence of FM intrachain interactions with interchain AF interactions, in good agreement with the analysis of the intermolecular contacts through the McConnell I model [44].

3.1.2 $[M(Cp^*)_2][Ni(edt)_2]$

The compounds $[M(Cp^*)_2][Ni(edt)_2]$, with M = Fe and Cr, are isostructural and the crystal structure [41] consist of a parallel arrangement of type I chains. In Fig. 9a a view along the chain direction ([1 0 1]) is presented for $[Fe(Cp^*)_2][Ni(edt)_2]$.

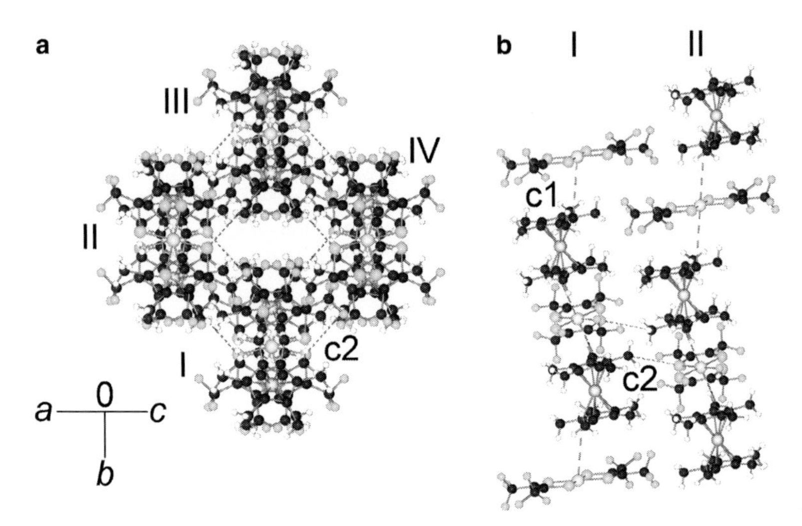

Fig. 8 (**a**) View of the crystal structure of $[Fe(Cp^*)_2][Ni(tdt)_2]$ along the chain direction. (**b**) Interchain arrangement of the pair I–II, c1 corresponds to the closest intrachain contact (D^+A^-) and c2 to the closest interchain contact (D^+A^-)

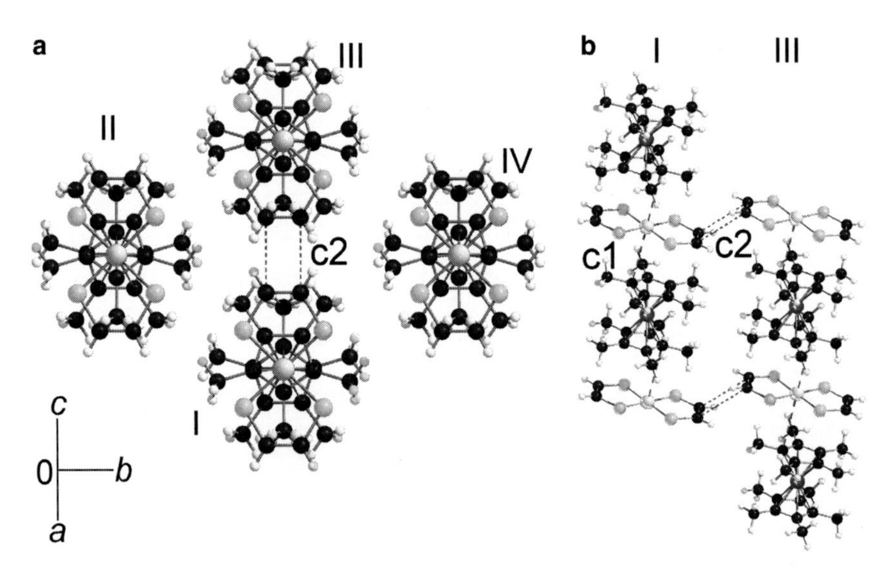

Fig. 9 (**a**) View of the crystal structure of $[Fe(Cp^*)_2][Ni(edt)_2]$ along the chain direction. (**b**) Interchain arrangement of the pair I–III, c1 corresponds to the closest intrachain contact (D^+A^-) and c2 to the closest interchain contact (A^-A^-)

Within the chains the Ni atoms sit above the Cp fragments from the cations, Ni...C D^+A^- contacts (c1) shorter than the sum of the van der Waals radii, d_W, were observed. For this compound the shortest interchain interionic separation was found in the in-registry pair I-IIII, with A^-A^- C...C contacts (c2), exceeding d_W by

11%, as shown in Fig. 9b. The pairs I–II and I–IV present a similar interchain arrangements, with D^+ A^- C. . .S contacts, exceeding d_W by 11%.

In the case of the $[M(Cp^*)_2][Ni(edt)_2]$ compounds, AF interactions are dominant at high temperatures, as seen by the negative θ values (Table 1). A considerable field dependence of the obtained θ values on the measurements of polycrystalline samples (free powder) was observed in the case of $[Fe(Cp^*)_2][Ni(edt)_2]$, which can be attributed to the magnetic anisotropy of the cation and to the anisotropic magnetic coupling in this compound [41]. This is consistent with the metamagnetic behavior observed at low temperatures, with $T_N = 4.2$ K and $H_C = 14$ kG at 2 K. The typical metamagnetic behavior was confirmed by single crystal magnetization measurements at 2 K [41], shown in Fig. 10. With the applied magnetic field parallel to the chains a field induced transition from an AF state to a high field "FM" state occurs at a critical field of 14 kG. A linear field dependence was observed for the magnetization, with the applied field perpendicular to the chains as expected for an AF and no field induced transition was detected.

The magnetic behavior of $[Fe(Cp^*)_2][Ni(edt)_2]$ is consistent with the coexistence of FM intrachain interactions, due to D^+A^- intrachain short contacts, with AF interchain interactions, resulting from the D^+A^- and A^-A^- interchain contacts, as predicted by the McConnell I mechanism [44]. In this case the interchain interactions must be particularly large as they seem to be the dominant interactions at high temperatures. The value of the critical field, 14 kG, much higher than any of the

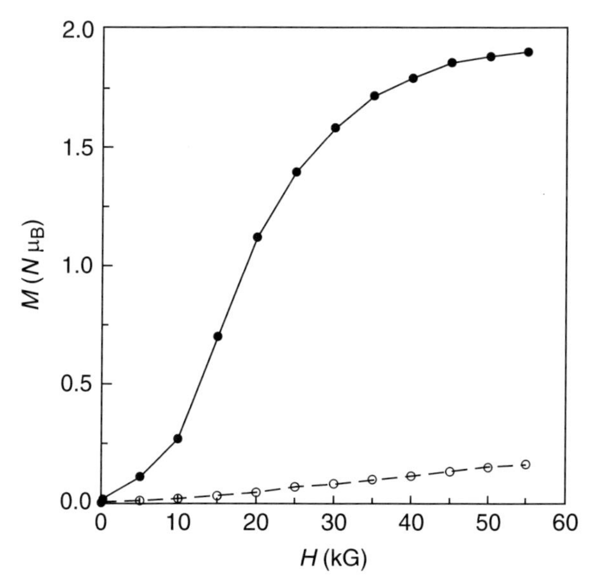

Fig. 10 Magnetization isothermal at 2 K of a $[Fe(Cp^*)_2][Ni(edt)_2]$ single crystal, the *closed symbols* refer to measurements with applied field parallel to the chains and the *open symbols* to the measurements with the applied field perpendicular to the chains

other of the metamagnets reported here, is consistent with the existence of quite strong AF interchain coupling.

3.1.3 $[Fe(C_5Me_4SCMe_3)_2][M(mnt)_2]$, M = Ni, Pt

The compounds $[Fe(C_5Me_4SCMe_3)_2][M(mnt)_2]$ (M = Ni, Pt) are the only cases of salts based on metallocenium derivatives and on $[M(mnt)_2]^-$ complexes where the crystal structure is based in parallel arrangements of the type I chain motif [49]. In the chains the $[Pt(mnt)_2]$- units are considerably tilted in relation to the chain direction, and short interatomic D^+A^- intrachain contacts were observed, involving one C from the Cp and a S atom from the anion, with a C-S distance of the order of d_W. Relatively short interchain interionic distances were observed, where the closest corresponds to a S...C, involving one S atom from the anion and a C atom from a Me group of the cation, exceeding d_W by 10%.

The magnetic behavior of $[Fe(C_5Me_4SCMe_3)_2][M(mnt)_2]$ (M = Ni, Pt) is dominated by FM interactions ($\theta = 3$ K, M = Ni and Pt), which can be attributed to the D^+A^- intrachain interactions. The weaker interchain interactions are expected to be AF. As in the previous compounds exhibiting this type of structure a metamagnetic behavior is also expected to occur at low temperatures.

3.1.4 β-$[Fe(Cp^*)_2][Pt(mnt)_2]$

The crystal structure of β-$[Fe(Cp^*)_2][Pt(mnt)_2]$ consists of parallel alternated $D^+A^-D^+A^-$ (type I) chains, which are isolated by chains of D^+ $[A_2]^{2-}D^+$ units [14]. The type I chains are considerably separated (16.576 Å). Within the chains both the $[Fe(Cp^*)_2]^+$ and $[Pt(mnt)_2]^-$ units are considerably tilted in relation to the chain direction, and the Cp ring sits on top of the ethylenic C=C of the mnt^{2-} ligands from the anions. Short interatomic D^+A^- intrachain distances were observed, involving one C from the Cp ring and a S atom from the anion, with a C-S distance exceeding d_W by 5%. In the D^+ $[A_2]^{2-}D^+$ units the anions are strongly dimerized through a Pt-Pt bond (3.574 Å).

The magnetic susceptibility of β-$[Fe(Cp^*)_2][Pt(mnt)_2]$ follow a Curie-Weiss behavior with $\theta = 9.8$ K. The dominant ferromagnetic interactions are assigned to the magnetic intrachain D^+A^- interactions from the type I chains, as the contribution from the D^+ $[A_2]^{2-}D^+$ unit chains is expected to respect only to the cations due to the strong dimerization of the anions, $S = 0$ for $[A_2]^{2-}$.

3.1.5 $[M(Cp^*)_2][M'(\alpha\text{-tpdt})_2]$

The compounds from this family $[M(Cp^*)_2][M'(\alpha\text{-tpdt})_2]$ with M = Co, Fe, Mn, Cr, and M′ = Ni, Au, have been reported and although not strictly isostructural they present a similar structure based on the same packing pattern [28, 47, 48]. This is an example of a family of compounds where by comparison of their magnetic properties it was possible to clearly put into evidence the role of the cation and anion magnetic moments and cation magnetic anisotropy in the magnetic properties of these compounds, and to probe in detail the nature of intermolecular magnetic interactions.

The crystal structures of the compounds of this family, in spite of being based on the type I structural chain motive, are considerably different from the structures presented before [47] and they can be seen as composed of alternated layers, each consisting of parallel alternated $D^+A^-D^+A^-D^+A^-$ chains packed in an out-of-registry mode, the chains of adjacent layers being nearly perpendicular (88.7°). A projection of the crystal structure of $[Cr(Cp^*)_2][Ni(\alpha\text{-tpdt})_2]$ along [1 −1 0] is shown in Fig. 11a. A view of two adjacent layers of chain, along the direction [1 0 2] is shown in Fig. 11b, where on the left the chains are aligned along [1 −1 0] and on the right along [1 1 0].

The D^+A^- intrachain contacts are made with the cation Cp rings sitting on the top of the thiophenic rings of the anions, with D^+A^- atomic contacts (c1), with separations exceeding slightly d_W (from ~2–10%), denoting strong π−π interactions. Due to the out-of-registry arrangement of the chains there are also relatively short D^+A^- contacts (c3) between neighboring chains, namely between an S atom from the MS_2C_2 fragment from the anion and a C atom from the Me groups in the cation. In addition there are contacts between the chains in adjacent layers, namely relatively short interanionic A^-A^- contacts (c2) between S atoms from the central NiS_4 fragment and a C atom from the thiophenic fragment of the ligand, and also C...C contacts through H atoms. Therefore this structure can also be seen as built from alternated layers of cations and anions, parallel to *b, c* as in the first description adopted for $[Fe(Cp^*)_2][Ni(\alpha\text{-tpdt})_2]$ [28], but the contacts in these anionic and cationic layers are less significant than along the D^+A^- chains.

The compound $[Fe(Cp^*)_2][Ni(\alpha\text{-tpdt})_2]$, although not isostructural with the $[Cr(Cp^*)_2]$ analogue, presents a lower symmetric structure, where one of the crystallographic axis is doubled. This axis doubling is associated with a slight alternation in the D^+–A^- contacts along the chains and consequently also in interchain contacts. Otherwise the network of intermolecular contacts remains similar. There are no full structural refinements for the other compounds of this series but the unit cell parameters of the (M = Fe, M′ = Au) compound and powder diffraction data of the salt with M = Co, M′ = Ni indicate that they are isostructural with the M = Fe, M′ = Ni compound. On the other hand, powder diffraction data on the M = Mn, M′ = Ni compound indicate that it is isostructural with the M = Cr, M′ = Ni compound.

The χ(T products for the [M(Cp*)2][Ni((-tpdt)2] salts with M = Fe, Mn and Cr, increase upon cooling indicating dominant ferromagnetic (FM) interactions in these compounds. The paramagnetic susceptibility above 30 K follows a Curie-Weiss behavior: χ(= C/(T - θ(), with θ (values of 3.8, 7.5, and 3.0 K for M = Fe, Mn and Cr respectively. This behavior contrasts with the analogues with diamagnetic cations [Co(Cp*)2] [Ni(α-tpdt)2] (SD = 0 and SA = 1/2) and with diamagnetic anions, [Fe(Cp*)2][Au(α-tpdt)2], (SD = 1/2 and $S_A = 0$) where a continuous decrease of the

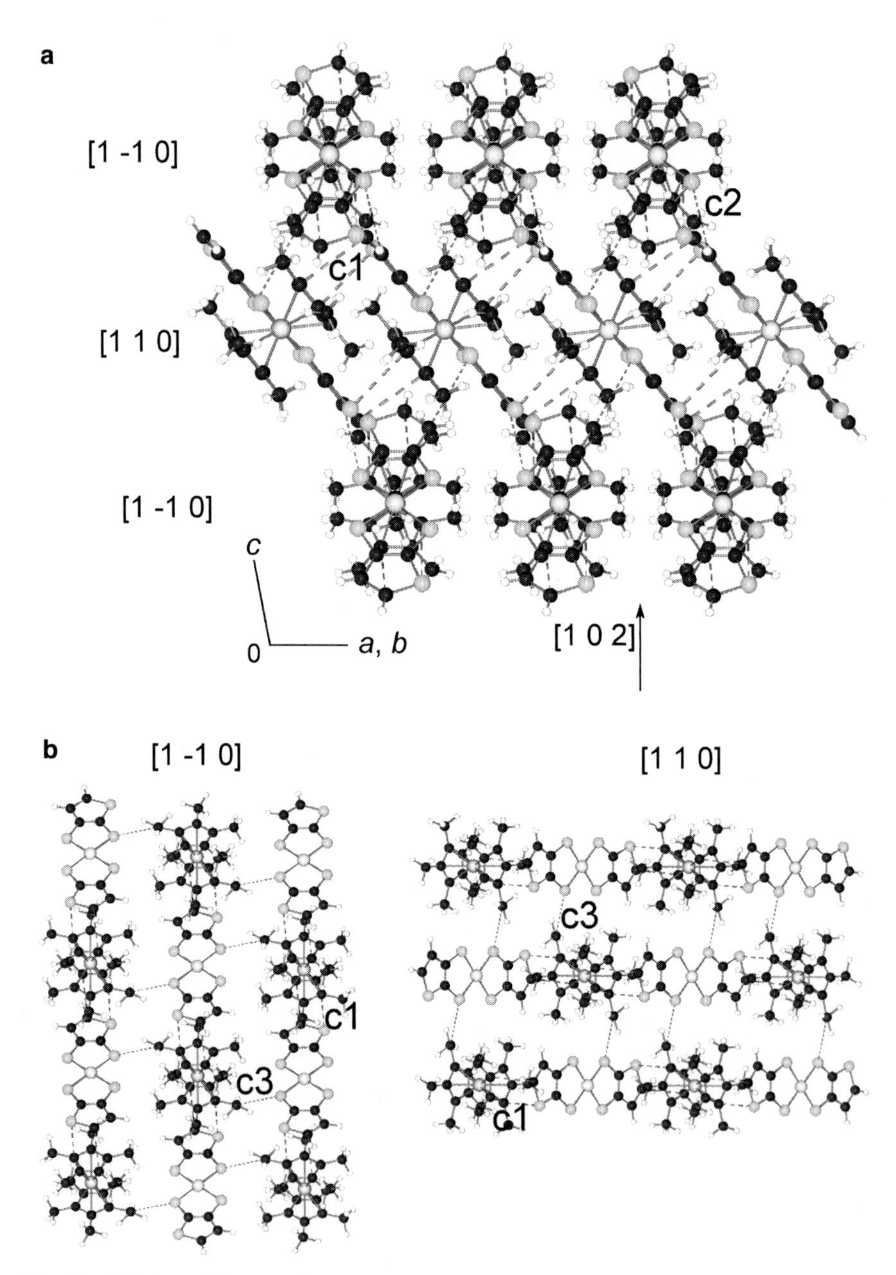

Fig. 11 **(a)** View of the crystal structure of $[Cr(Cp^*)_2][Ni(\alpha\text{-tpdt})_2]$ along [1 −1 0], where it is possible to see three layers of chains aligned either along [1 −1 0] or [1 1 0], c1 correspond to the closest D^+A^- intrachain contacts and c2 to the closest A^-A^- interchain interlayer contacts. **(b)** View of two chain layers along the [1 0 2] direction. The chains of the layer on the right are aligned along [1 −1 0] and on the left along [1 1 0]. c3 corresponds to the closer interchain contacts within the layers

χT product is observed upon cooling. This indicates that both the anion-anion and cation-cation interactions in these compounds are antiferromagnetic and that the dominant ferromagnetic interactions in the salts with both paramagnetic cations and anions are ascribed to intrachain cation-anion interactions. These findings are in agreement with the predictions of the McConnell I model for the magnetic interactions based on the spin density calculations that predicted FM (D^+A^-) intrachain coupling coexisting with weaker AF interchain coupling (D^+A^- intra and A^-A^- interlayer) [47]. It should be noted in this respect that in the anion $[Ni(\alpha\text{-tpdt})_2]^-$, at variance with the previously mentioned small anions, there is a significant spin polarization effect with small negative spin densities in the S atom and in one of the carbons from the thiophenic ring.

In spite of the common dominant ferromagnetic interactions in the compounds $[M(Cp^*)_2][Ni(\alpha\text{-tpdt})_2]$, M = Fe, Mn and Cr, their low temperature behaviors are different. The salt of the $[Cr(Cp^*)_2]^+$ ($S = 3/2$) isotropic cation, remains paramagnetic with no phase transition down to the lowest temperatures measured (1.6 K) [47]. In [Mn(Cp*)2][Ni(α-tpdt)2], where the ($S_D = 1$) cation exhibits a significant single ion magnetic anisotropy, in agreement with a.c. susceptibility measurements, a magnetic behavior typical of a frustrated magnet is observed, with a blocking temperature of the order of 4 K. The magnetic frustration in this compound is ascribed to a degenerate ground state in the interlayer spin arrangement.

The magnetization of $[Fe(Cp^*)_2][Ni(\alpha\text{-tpdt})_2]$ under low magnetic fields shows a maximum upon cooling corresponding to a transition to an AF ground state with a critical temperature $T_N = 2.56$ K under zero field. However, the application of a magnetic field above 800 G completely suppresses this transition. At low temperatures, below $T_N = 2.56$ K, the magnetization shows a metamagnetic behavior with $H_C = 600$ G, at 1.7 K; the magnetization isothermals exhibit the typical sigmoidal behavior and above 7 T magnetization becomes almost saturated, reaching values close to saturation attaining 2.45 μ_B mol^{-1} at 12 T, a value quite close to that predicted with $g_{Ni} = 2.07$ and $g_{Fe} = 2.81$. The temperature-magnetic field (T, H) phase diagram of this compound was investigated by a combination of a.c. susceptibility and d.c. magnetization measurements as a function of temperature and magnetic field. In addition to the low temperature-low magnetic field AF phase, and the low temperature-high field aligned paramagnetic region (FAP), there is evidence for an intermediate phase (I) of nature but this is not yet clear (Fig. 12). Also in this phase diagram the field aligned paramagnetic region appears separated from the high temperature paramagnetic phase (P) by an unclear boundary, dashed line in Fig. 12, denoted by broad peaks in a.c. susceptibility, which however were not clearly ascribed to a real phase transition.

The metamagnetic behavior of $[Fe(Cp^*)_2][Ni(\alpha\text{-tpdt})_2]$ is attributed to the AF coupling between the FM coupled $D^+A^-D^+A^-$ chains within the chain layers, as predicted from the application of the McConnell I model and spin density calculations, in a similar way to other salts also based on decamethylmetallocenes and other transition-metal bisdichalcogenate complexes with a type I structural motive such as $[Mn(Cp^*)_2][M(tdt)_2]$ (M = Ni, Pd, Pt).

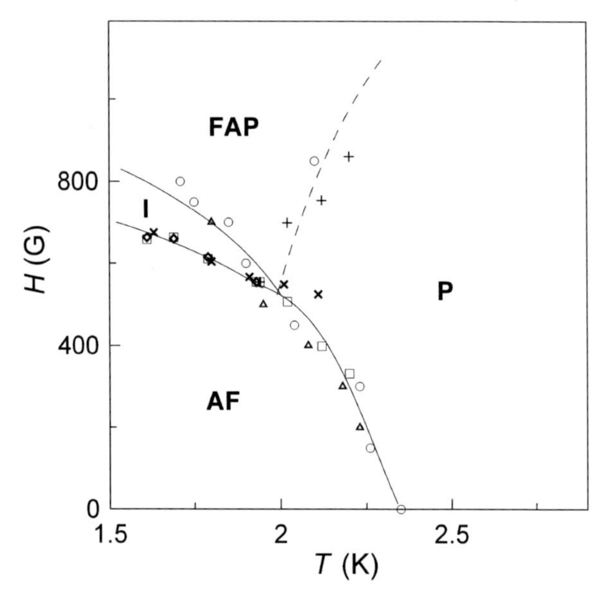

Fig. 12 Magnetic phase diagram of $[Fe(Cp^*)_2][Ni(\alpha\text{-tpdt})_2]$. Boundaries detected by different measurements: *M(T)* (*open triangles*); *M(H)* (*crosses*); $\chi(T)$(*open circles*) and $\chi''(H)$ (*open squares* and *plus signs*)

A detailed analysis of the magnetization curves of $[Fe(Cp^*)_2][Ni(\alpha\text{-tpdt})_2]$ reveals an unusual hysteresis with a partially inverted loop with a crossover between the increasing and decreasing field curves at ~1 kG and a negative coercivity in the field induced transition region (~0.65 kG) as shown in Fig. 13 [73]. For loops between 0 and 3 kG, the coercivity in the field induced transition is of the order of −20 G, but it increases significantly with higher applied magnetic fields and can reach values of −100 G. This is the first molecular material where negative coercivity was observed and its origin remains not completely clear but it has to be related to the inversion with the applied magnetic field of the cations spins, which exhibit a considerable magnetic anisotropy, and are AF coupled between neighboring layers. A somewhat similar inverted hysteresis behavior was also observed in the $[Mn(Cp^*)_2][Ni(\alpha\text{-tpdt})_2]$ compound below 4 K [73], as shown in Fig. 14. The absence of a clear metamagnetic, as observed in $[Fe(Cp^*)_2]$ $[Ni(\alpha\text{-tpdt})_2]$, can be attributed to a significant decrease of the AF interchain intralayer coupling or to a change in the nature of the interchain magnetic coupling.

3.2 *Type II Mixed Chain $[M(Cp^*)_2][M'(L)_2]$ Salts*

This type of chain arrangement was observed only in the case of α-$[Fe(Cp^*)_2][Pt(mnt)_2]$. In the crystal structure of this salt, layers of parallel type II chains, with a net charge (−) per repeat unit, $[A_2]^{2-}D^+$, alternate with layers presenting a D^+D^+ A^-

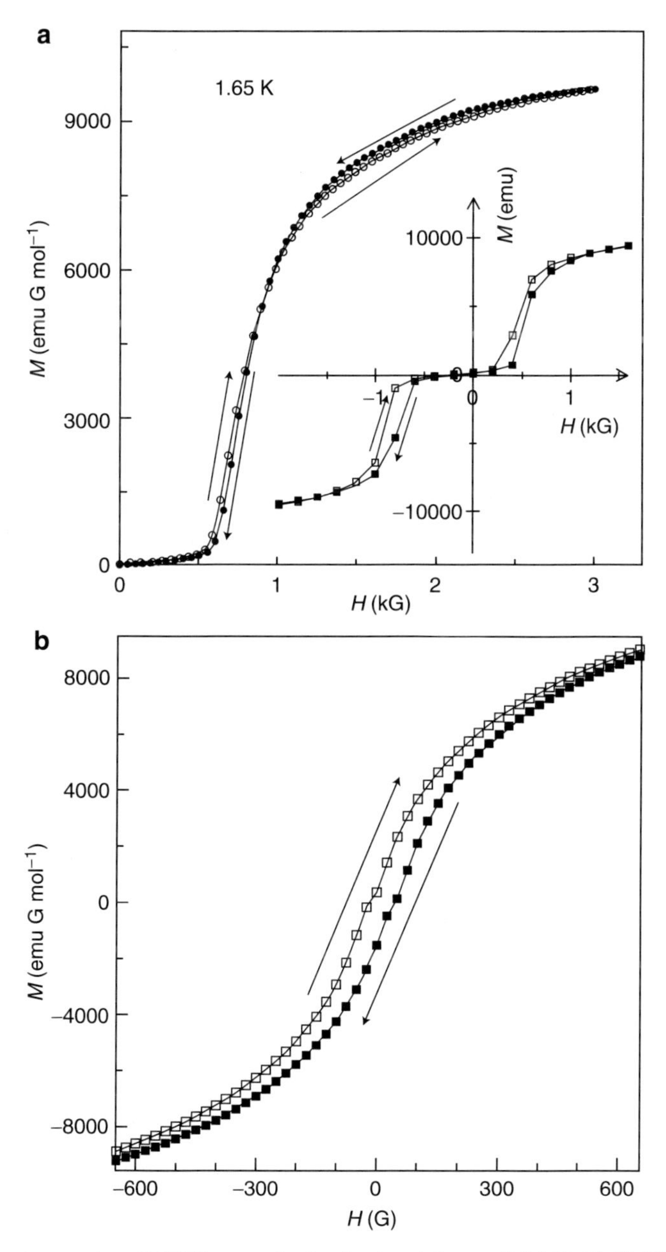

Fig. 13 Low temperature (1.65 K) magnetization hysteresis loops of: (**a**) $[Fe(Cp^*)_2][Ni(\alpha\text{-tpdt})_2]$ and (**b**) $[Mn(Cp^*)_2][Ni(\alpha\text{-tpdt})_2]$

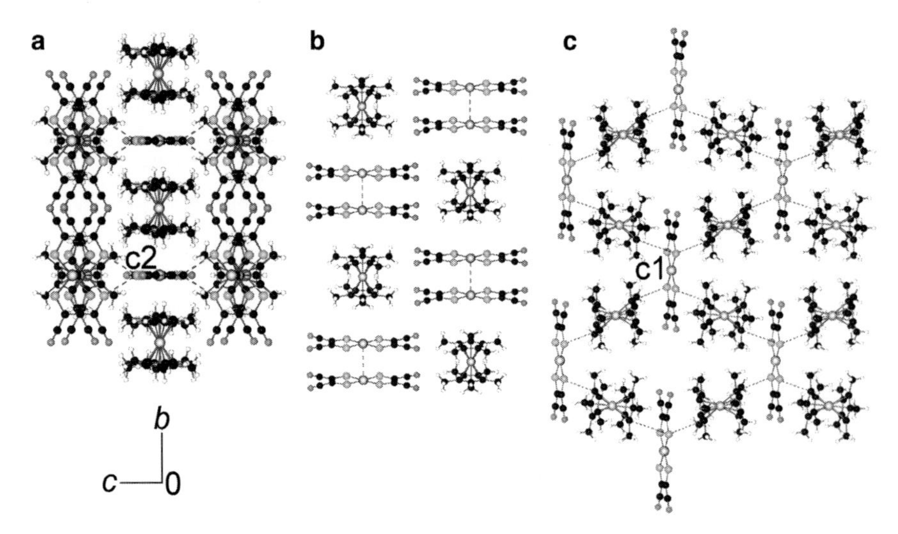

Fig. 14 (**a**) View of the structure of α-[Fe(Cp*)$_2$][Pt(mnt)$_2$] along the type II chains direction. (**b**) View of a type II chain layer. (**c**) View of the $D^+D^+A^-$ layer. c1 corresponds to the closest contact in these layers and c2 to the closer contacts between the two types of layers (A^-A^-)

repeating unit, with a net (+) per repeat unit [14]. A view of the crystal structure along the chain directions, [1 0 0], is shown in Fig. 14a representing two type II chain layers (left and right) and in between the D^+ D^+ A^- layer. In the type II chain layers the anions are strongly dimerized through a Pt-Pt bond of 3.575 Å, and the $[A_2]^{2-}D^+$ $[A_2]^{2-}D^+$ chains present an out-of-registry arrangement, as shown in Fig. 14b. Apart the A^-A^- (Pt-Pt) contacts no other short contacts were observed in these layers. The D^+ D^+ A^- layer presents a unique arrangement, where the cations sit on top of the extremity of the anions, these D^+ D^+ A^- units form edge to edge chains, as shown in Fig. 14c. In these layers the closest interionic separations (c1) involve one C from the Cp ring and an S atom from the anion, with a C...S separation exceeding d_W by ~15%. Relatively close interlayer A^-A^- (S...S) contacts (c2) where detected, exceeding d_W by around 8%.

The magnetic susceptibility of α-[Fe(Cp*)$_2$][Pt(mnt)$_2$] follows a Curie-Weiss behavior with $\theta = 6.6$ K [14]. The dominant FM interactions are assigned to the magnetic D^+ A^- interactions from the D^+ D^+ A^- layers, as the contribution from the $[A_2]^{2-}D^+$ chains is expected with respect only to the isolated cations due to the strong dimerization of the anions, $S = 0$ for $[A_2]^{2-}$.

3.3 *Type III Mixed Chain [M(Cp*)$_2$][M′(L)$_2$] Salts*

The salts exhibiting this type of basic structural motive are based on metal bisdithiolene complexes with dmit type ligands (dmit, dmio or dsit). These complexes present an extended π system with a large number of heteroatoms (S, Se and O) in the periphery and in the solid state the anionic complexes are frequently associated as either weakly or strongly dimerized pairs, as it occurs in the type III structural motive. As described below, the magnetic behavior of the [M(Cp*)$_2$][M′(L)$_2$] salts based on type III chains, is strongly dependent on the arrangements in the face-to-face pairs of anions.

3.3.1 [M(Cp*)$_2$][M′(dmit)$_2$] (M = Fe, and M′ = Ni, Pd, Pt; M = Mn, and M′ = Ni, Au; M = Co and M′ = Ni) and β-[Fe(Cp*)$_2$][Pd(dmit)$_2$]

The salts [Fe(Cp*)$_2$][M(dmit)$_2$] (M = Ni [15] and Pt [55, 56]), [Mn(Cp*)$_2$][M(dmit)$_2$] (M = Ni [52] and Au [52]) and [Co(Cp*)$_2$][Ni(dmit)$_2$] [54] are isostructural and their crystal structure consists of 2D layers composed of parallel type III chains. A view of the structure along the chains direction is shown in Fig. 15a for [Fe(Cp*)$_2$][Ni(dmit)$_2$]. Within the chains the Cp fragments of the cation sit above the dmit ligands of the anion, as illustrated in Fig. 15b. Intrachain short contacts between the [Ni(dmit)$_2$]$^-$ units in the anion pairs (c1), involving a Ni atom and a S atoms from the five membered ring C_2S_2C from the ligand. The intrachain D^+ A^- contacts (c2) present larger separations, and the shorter D^+ A^- contacts involve C atoms of the Cp rings from the cations and S atoms from the C_2S_2C ring of the anions ligands, with distances exceeding d_W by ~5%. A variety of S...S short interchain contacts (c3) between the [Ni(dmit)$_2$]$^-$ units were observed, with shorter separations than c1, connecting the chains within the layers anion molecules, as shown in Fig. 15b, and even between chains in adjacent layers (c4), as shown in Fig. 15a and b (dashed lines). This network of S...S (and Ni...S) short contacts gives rise to an anionic 3D network, with cavities accommodating the pairs of cations.

In spite of the similarities in the crystal structures of the compounds [Fe(Cp*)$_2$][M(dmit)$_2$], with M = Ni and Pt, and [Mn(Cp*)$_2$][Ni(dmit)$_2$], where S= 1/2 for the anions and S= 1/2 for [Fe(Cp*)$_2$]$^+$ and S= 1 for [Mn(Cp*)$_2$]$^+$, they present distinct magnetic behaviors. In the case of [Fe(Cp*)$_2$][Ni(dmit)$_2$], a minimum in the temperature dependence of χT, is observed at 30 K [51], as shown in Fig. 16 (squares). This could be ascribed either to a change of the dominant magnetic interactions, due to structural changes with cooling, or to a ferrimagnetic (FIM) behavior, which seems more likely to occur, considering the magnetic behavior exhibit by [Mn(Cp*)$_2$][Ni(dmit)$_2$]. At high temperatures the magnetic susceptibility follows a Curie-Weiss behavior, with a θ value of -7.6 K, which clearly indicates that AF interactions are dominant. However, below $T_m = 30$ K, χT increases rapidly indicating that apparently the FM interactions become dominant in that region. This is further confirmed by the magnetization field dependence at low temperatures (squares) shown in Fig. 17, which for low applied magnetic fields increases faster than predicted by the Brillouin function (solid line).

In the case of [Mn(Cp*)$_2$][Ni(dmit)$_2$], at high temperatures FM interactions are dominant (θ = 2.5 K), and an FIM transition was reported to occur at 2.5 K [52]. The field cooled (FCM), zero field cooled (ZFCM) and remnant (REM) magnetization temperature dependencies are shown in Fig. 18. This is the first

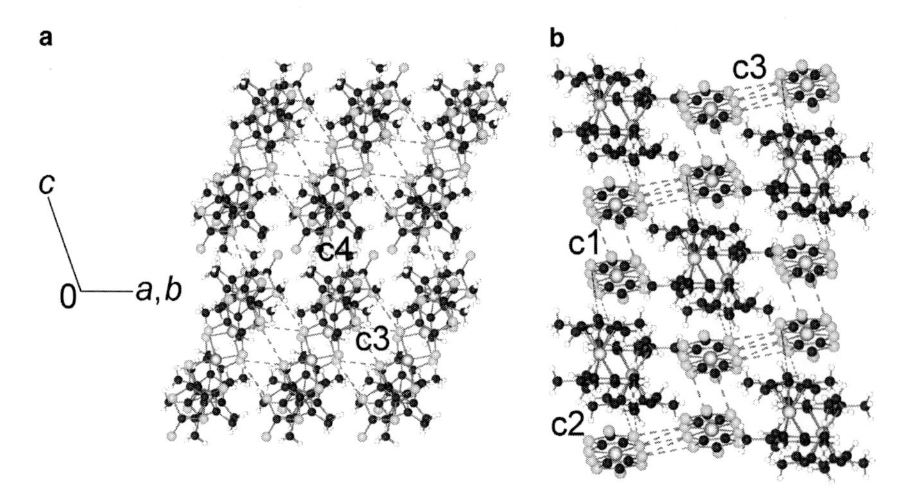

Fig. 15 (**a**) View of the crystal structure of $[Fe(Cp^*)_2][Ni(dmit)_2]$ along the stacking direction. (**b**) View of one layer of the type II chains. c1 corresponds to the intradimer (AA) closest contact, c2 to the closer D^+A^- intrachain contacts, c3 include various S-S short A^-A^- interchain contacts, and c4 concerns a short S-S interlayer (A^-A^-) contact

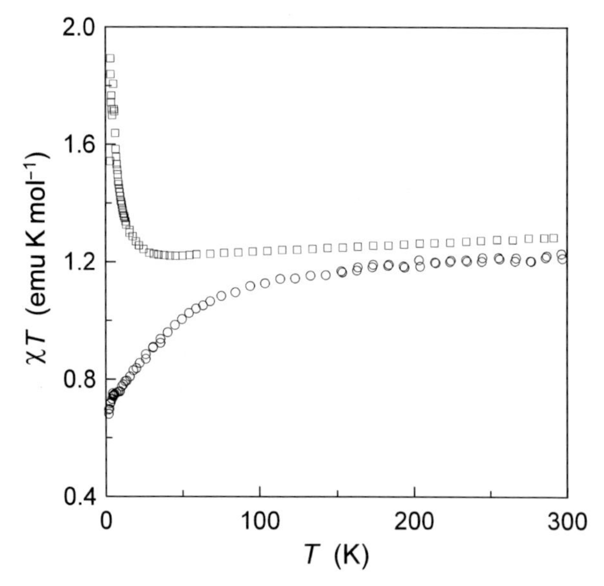

Fig. 16 χT temperature dependence for $[Fe(Cp^*)_2][Ni(dmit)_2]$ (*squares*) and $[Fe(Cp^*)_2][Pt(dmit)_2]$ (*circles*)

report of an FIM ordering among decamethylmetallocenium-based salts and the magnetic behavior of $[Mn(Cp^*)_2][Ni(dmit)_2]$ was analyzed in the framework of the McConnell I mechanism [52]. The $[Ni(dmit)_2]^-$ anions were found to present a spin polarization effect and according to the McConnell model the interactions between

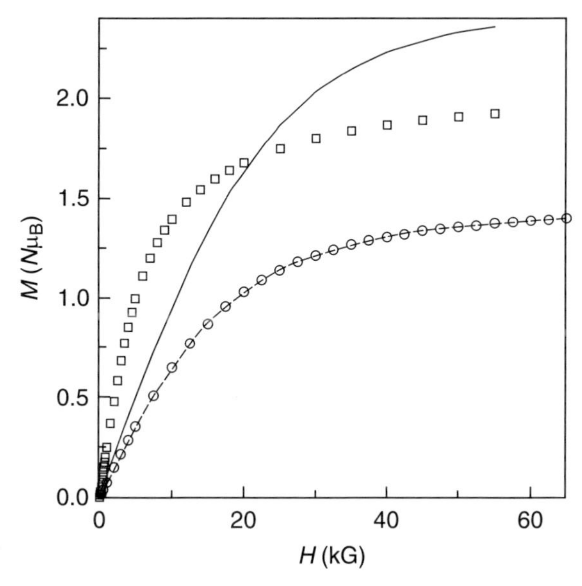

Fig. 17 Magnetization isothermals of $[Fe(Cp^*)_2][Ni(dmit)_2]$ (*squares*) and $[Fe(Cp^*)_2][Pt(dmit)_2]$ (*circles*) at 1.8 K. The *solid line* corresponds to calculated values from the Brillouin function, while the *dashed line* corresponds only the contribution from the cation molecules. From [53]

the $[Ni(dmit)_2]^-$ units in the anion pairs are predicted to be FM, as the atoms involved in the intradimer contacts present different signs in the atomic spin density (the Ni atom presents a positive spin density while in the S atoms from the C_2S_2C ligands fragments, the spin density is negative) [52]. The origin of the FIM ordering occurring at $T_N = 2.5$ K is then attributed to AF intrachain coupling between the cations and anions [52]. The isothermal obtained at 2 K for this compound shows a dramatic increase at low fields attaining a value of 1.3 μ_B at 200 G, slightly above the calculated value for the FIM state (1.17 μ_B), and above this field it increases slowly, as shown in Fig. 19. This small discrepancy is attributed to the anisotropy of the g value of the $[Mn(Cp^*)_2]^+$ cation and to the proximity of the transition. A small coercive field typical of a magnet, close to the transition temperature, of ~3.5 G was detected at 2 K, as shown in Fig. 20. The analysis of the crystal structure reveals the existence of a large number of intrachain short contacts between the anions, which are expected to lead to sizable AF interactions between the chains. In spite of these contacts being quite short they do not seem to play a significant role in the magnetic behavior of the $[Mn(Cp^*)_2][Ni(dmit)_2]$, which can be attributed to a rather weak overlap associated with those contacts.

In the case of the $[Mn(Cp^*)_2][Au(dmit)_2]$ salt, with the $[Au(dmit)_2]^-$ ($S = 0$) diamagnetic anions, the paramagnetic susceptibility is due only to the contributions of the $[Mn(Cp^*)_2]^+ S = 1$ cations and it follows a Curie-Weiss behavior, with a θ value of -4.2 K [52]. The AF interactions in this compound are assigned to the weak AF coupling between the cations.

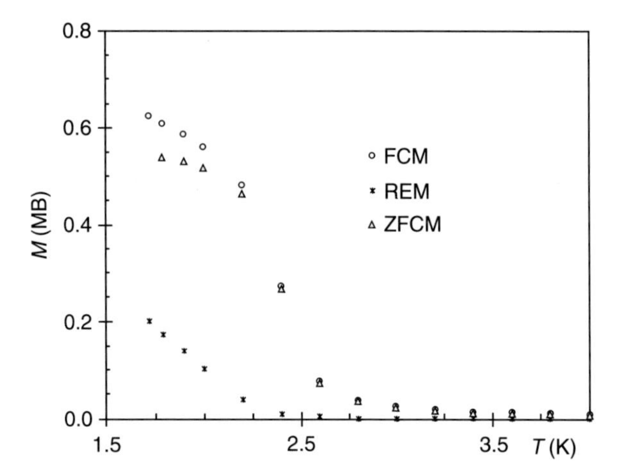

Fig. 18 Field cooled (FCM), zero field cooled (ZFCM) and remnant (REM) magnetization temperature dependencies of $[Mn(Cp^*)_2][Ni(dsit)_2]$. From [53]

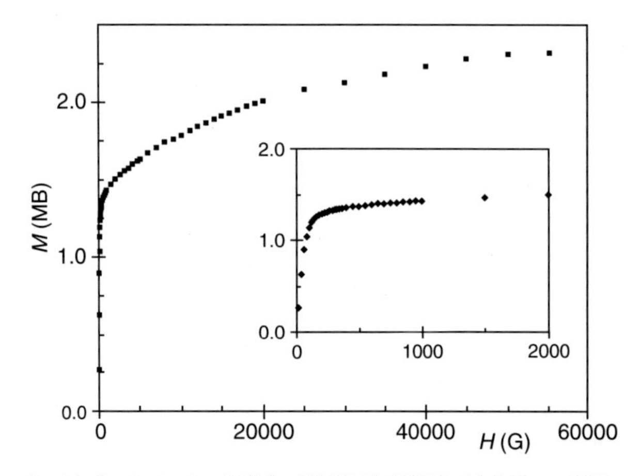

Fig. 19 Magnetization isothermal at 2 K for $[Fe(Cp^*)_2][Ni(dmit)_2]$. From [52]

In $[Co(Cp^*)_2][Ni(dmit)_2]$, where the $[Co(Cp^*)_2]^+$ cations are diamagnetic ($S = 0$), the paramagnetic susceptibility arises only from the contribution of the anionic network, where $S = 1/2$ for $[Ni(dmit)_2]^-$. The magnetic behavior of this compound is dominated by weak FM interactions ($\theta = 0.5$ K) [54]. The observed dominance of the FM interactions in this compound confirms that the FM interactions between the $[Ni(dmit)_2]^-$ units in the dimers are stronger than the AF A^-A^- intrachain interactions.

Unlike that observed in $[M(Cp^*)_2][Ni(dmit)_2]$, with M = Fe and Mn, in $[Fe(Cp^*)_2][Pt(dmit)_2]$ the χT product decreases significantly upon cooling as shown in Fig. 16 (circles), and the magnetic behavior is dominated by strong AF interactions.

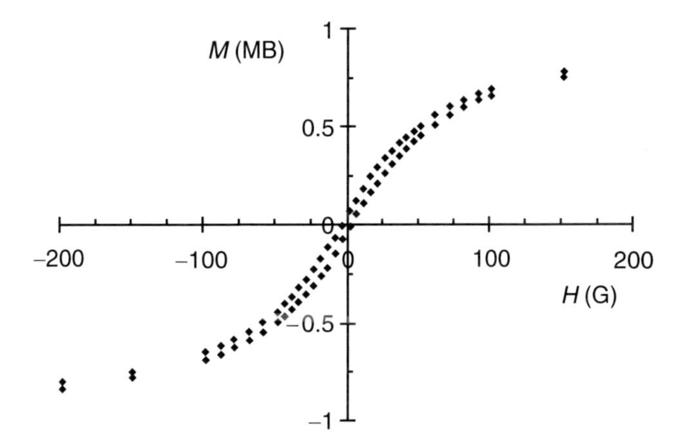

Fig. 20 Magnetization hysteresis cycle at 2 K for $[Fe(Cp^*)_2][Ni(dmit)_2]$. From [52]

From a poor fit to a Curie-Weiss law a Weiss constant, θ, of -14.4 K [51] was obtained. For this compound the magnetization isothermals (circles in Fig. 17) at low temperatures ($T < 2$ K) are consistent with a total canceling of the paramagnetic contribution from the anions, and the magnetization field dependence follows the calculated Brillouin function for the isolated cations, $[Fe(Cp^*)_2]^+$, spins (dashed line) as shown in Fig. 16. The χT temperature dependence fitted quite well to a sum of two contributions, a Curie law to account for the contribution of the cations and a Bleaney-Bowers equation to account for the anions contribution, with $2J/k = -112$ K. In spite of this compound being isostructural with the salts $[M(Cp^*)_2][Ni(dmit)_2]$, with M = Fe and Mn, and a similar spin polarization effect being expected in the $[Pt(dmit)_2]^-$ anion, the experimental data show that the coupling between the anions in the dimer is AF in the case of $[Fe(Cp^*)_2][Pt(dmit)_2]$. This can be attributed to a subtle difference in the overlap between the two anionic units in the pair leading to an AF coupling in this compound, as, in case of the $[Ni(dmit)_2]^-$ pairs, a competition between FM and AF interactions was observed [52]. The overlaps between the $[Ni(dmit)_2]^-$ and $[Pt(dmit)_2]^-$ pairs are compared in Fig. 21. In the $[Ni(dmit)_2]^-$ anions, with the exception of the S atoms (dark atoms) from the outer C_2S_2C ring that have a negative spin density, all the other atoms show positive spin densities. As shown in Fig. 21, the two pairs of anions contacts between atoms with the same sign of the spin density (AF interactions) are represented by the clear areas (i and iii), while the contacts involving atoms with opposite signs of the spin density (FM coupling) are denoted by the darker regions (ii). In case of $[Ni(dmit)_2]^-$, in spite the larger number of AF contacts, the FM contact is expected to prevail as most of the spin density resides on the central NiS_4 core of the anion, and the contact ii is considerably shorter than the others (i and iii). The main difference in the $[Pt(dmit)_2]^-$ pair resides in the contact iii, where a S…S contact is also involved, and in this case the spin density in these coordinating S atoms, is significantly larger than in the S atoms from the outer C_2S_2C ring. Thus an AF coupling is expected in this pair.

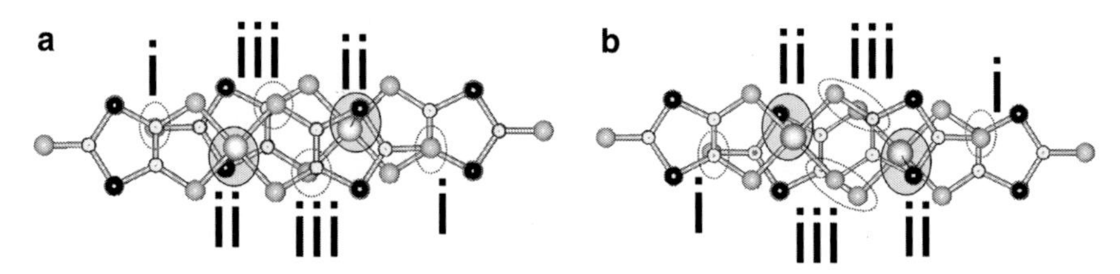

Fig. 21 The overlaps between the $[Ni(dmit)_2]^-$ (*left*) and $[Pt(dmit)_2]^-$ (*right*) pairs in $[Fe(Cp^*)_2][Pt(dmit)_2]$

In the case of $[Fe(Cp^*)_2][Pd(dmit)_2]$, two polymorphs were obtained, but the structure was only determined for the α phase, which is distinct from the above-mentioned compounds and will be described further down. In spite of the fact that the crystal structure of the β-$[Fe(Cp^*)_2][Pd(dmit)_2]$ could not be determined, this compound shows a magnetic behavior that resembles that observed for $[M(Cp^*)_2][Ni(dmit)_2]$, with M = Fe and Mn, exhibiting a ϑ value of 2.6 K [51]. Furthermore, the isothermal obtained at 2 K closely follows that exhibited by $[Ni(Cp^*)_2][Ni(dmit)_2]$. This strongly suggests that β-$[Fe(Cp^*)_2][Pd(dmit)_2]$ presents a similar structure as those compounds.

3.3.2 $[Fe(Cp^*)_2][M(dmio)_2]$ (M = Ni, Pd and Pt), α-$[Fe(Cp^*)_2][Pd(dmit)_2]$, $[Fe(Cp^*)_2][Ni(dsit)_2]$ and $[Mn(Cp^*)_2]_2[Ni(dmio)_2]_2{\cdot}PhCN$

The compounds $[Fe(Cp^*)_2][M(dmio)_2]$, with M = Pd and Pt, are isostructural and the crystal structure consists of a parallel arrangement of type III chains. The solid state structure is similar to the one observed in $[Fe(Cp^*)_2][Ni(dmit)_2]$, which was described in the previous section (Sect. 3.3.1). However for $[Fe(Cp^*)_2][M(dmio)_2]$ (M = Ni and Pt) the dimers show a different configuration. In this case no significant slippage is observed in the dimers and a quite short Pd^- Pd contact of 3.481 Å was observed in the case of $[Fe(Cp^*)_2][Pd(dmio)_2]$ [55, 56]. Also short intralayer A^-A^- contacts were detected, involving a S atom from the central MS_4 fragment and one of the S from the C_2S_2C fragment, with a S–S distance of 3.669 Å. No interlayer short contacts were observed for $[Fe(Cp^*)_2][Pd(dmio)_2]$. A similar dimeric arrangement is observed in $[Fe(Cp^*)_2][Pt(dmio)_2]$ [55, 56]. The absence of good quality single crystals has prevented so far the determination of the crystal structure of $[Fe(Cp^*)_2][Ni(dmio)_2]$. However the similar magnetic behavior, described below suggests that this compound presents a structure similar to the one described for $[Fe(Cp^*)_2][M(dmio)_2]$ (M = Ni and Pt).

The compound α-$[Fe(Cp^*)_2][Pd(dmit)_2]$ is not isostructural with $[Fe(Cp^*)_2][M(dmio)_2]$, with M = Pd and Pt, but presents a similar molecular arrangement in the crystal structure [55, 56]. Again, in the case of α-$[Fe(Cp^*)_2][Pd(dmit)_2]$, a strong dimerization was observed for the anions, presenting also very short Pd^- Pd contacts of 3.485 Å. Short intralayer contacts, similar to the observed in the previous compound, were observed, with a S-S distance of 3.558 Å. No interlayer short contacts were observed for α-$[Fe(Cp^*)_2][Pd(dmit)_2]$.

The crystal structure of $[Fe(Cp^*)_2][Ni(dsit)_2]$ is similar to those described previously in this section. In this salt the anions are strongly dimerized through Ni-Se bonds, the $[Ni(dsit)_2]^-$ units being slipped in such a way that the metal adopts a square pyramidal conformation, with an apical Ni–Se bond of 2.557 Å, which is only slightly larger than the average equatorial Ni–Se bond distance, 2.331 Å [60]. Short intralayer contacts, were observed, with a Se-Se distance of 3.562 Å. Short interchain interlayer contacts (S…S) were observed for $[Fe(Cp^*)_2][Ni(dsit)_2]$.

The crystal structure of $[Mn(Cp^*)_2]_2[Ni(dmio)_2]_2 \cdot PhCN$ [54], apart the presence of the solvent PhCN, is similar to that described for $[Fe(Cp^*)_2][M(dmio)_2]$ (M = Pd and Pt) and α-$[Fe(Cp^*)_2][Pd(dmit)_2]$, also showing an eclipsed arrangement in the pairs of the $[Ni(dmio)_2]^-$ units. However, in this compound a larger separation between the anionic units is observed, where the values of the average S…S intradimer contacts is 3.634 Å and the Ni…Ni separation is 3.615 Å. These values are larger than those observed for the other compounds, where S…S average separations are 3.503, 3.550 and 3.514 Å and the M…M are 3.480, 3.549 and 3.485 Å for $[Fe(Cp^*)_2][M(dmio)_2]$ (M = Pd and Pt) and α-$[Fe(Cp^*)_2][Pd(dmit)_2]$ respectively. Unlike these compounds, in the case of $[Mn(Cp^*)_2]_2[Ni(dmio)_2]_2{\cdot}PhCN$ no short intrachain A^-A^- contacts were observed.

The temperature dependence of the magnetic susceptibility of $[Fe(Cp^*)_2]$ $[M(dmio)_2]$ (M = Ni, Pd and Pt), α-$[Fe(Cp^*)_2][Pd(dmit)_2]$ and $[Fe(Cp^*)_2][Ni$ $(dsit)_2]$ is poorly described by a Curie-Weiss behavior, with θ values of -19.0, -24.7, -33.3, -22.3 and -18.9 K respectively, indicating that the magnetic behavior is clearly dominated by AF interactions. At low temperatures, the magnetic field dependence follows the predicted values for the isolated cations. As in the case of $[Fe(Cp^*)_2][Pt(dmit)_2]$, this suggest that the dimers present a singlet ground state due to the strong AF coupling between the $[M(L)_2]^-$ units. The magnetic behavior of $[Mn(Cp^*)_2]_2[Ni(dmio)_2]_2 \cdot PhCN$ was reported to follow a Curie-Weiss behavior, with a θ value of -3.62 K [54], considerably lower than the other compounds, which can be attributed to a considerably weaker AF coupling in the anion dimers. However, the χT low temperature limit is of the same order as that observed in the other compounds, and can be assigned to the contribution of the non-interacting cations.

As described before for $[Fe(Cp^*)_2][Pt(dmit)_2]$, in the case of $[Fe(Cp^*)_2][Ni$ $(dsit)_2]$ the χT temperature dependence showed a good agreement to an expression considering a sum of two contributions, a Curie term accounting for the contribution of the cations and a Bleaney–Bowers term accounting the anions contribution, with $2J/k = -95.8$ K [60]. As in those two compounds, a reasonable agreement with such an expression was also obtained for $[Fe(Cp^*)_2][M(dmio)_2]$ (M = Ni, Pd and Pt) and for α-$[Fe(Cp^*)_2][Pd(dmit)_2]$, where $2J/k$ = 114.6, 170.8, 182.6 and 172.0 K respectively.

The experimental behavior observed for these compounds shows good agreement with the analysis of the intermolecular compounds in the framework of the McConnell I model. For these compounds the intradimer AF interactions

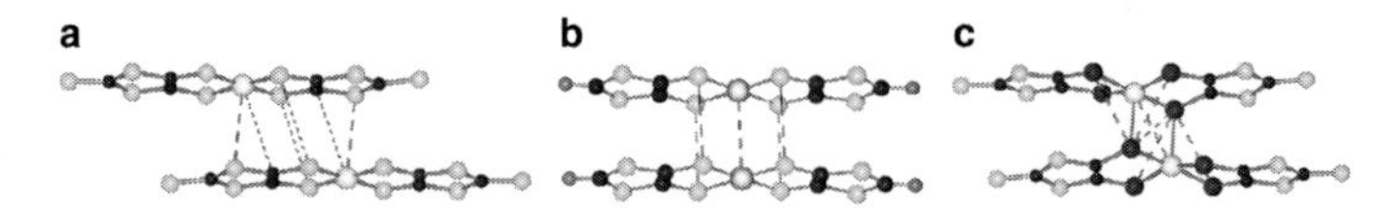

Fig. 22 The dimer arrangements of $[Fe(Cp^*)_2][Ni(dmit)_2]$ (*left*), $[Fe(Cp^*)_2][Pd(dmio)_2]$ (*center*) and $[Fe(Cp^*)_2][Ni(dsit)_2]$ (*right*)

are expected to be quite strong and dominate the magnetic behavior of these compounds.

The supramolecular arrangement in the type III chain based salts, at high temperatures, is consistent with the existence of dominant AF interactions through the A^-A^- interactions. For the compounds with dominant FM interactions at low temperatures, the magnetic behavior can be related to the distinct dimer arrangements in the pairs of the $[Ni(dmit)_2]^-$ units. The dimer arrangements along with the dimer overlapping are illustrated in Fig. 22 for $[Fe(Cp^*)_2][Ni(dmit)_2]$, $[Fe(Cp^*)_2]$ $[Pd(dmio)_2]$ and $[Fe(Cp^*)_2][Ni(dsit)_2]$. As referred to before, the contacts in the case of $[Fe(Cp^*)_2][Ni(dmit)_2]$, in spite of the competition between FM and AF contacts, are expected to lead to an overall FM intradimer coupling (unlike in $[Fe(Cp^*)_2][Pt(dmit)_2]$), while in case of the other two compounds this coupling is expected to be strongly AF.

3.4 *Type IV Mixed Chain [M(Cp*)₂][M′(L)₂] Salts*

The crystal structure of the salts based on type IV mixed chains consists of layers composed by parallel arrangements of those chains. In most cases the side-by-side pairs of cations alternate with monoanions, $\cdot\cdot A^-D^+D^+A^-D^+D^+\cdot\cdot$, and since there is a net charge (+) per repeat unit, $D^+D^+A^-$, the charge is compensated by anions in layers. As referred to previously, within the chains the pairs of cations present distinct arrangements, giving rise to three distinct subtypes of this structural motive. In the case of type IVa chains, both cations are aligned along the chains with the Cp* ligands parallel to the anions. In type IVb chains one of the cations is aligned with the chain, while the second is rotated by $\sim 90°$. Finally, in type IVc chains both cations are rotated by $\sim 90°$ relative to the chain axis, with the cation fivefold axis aligned along the long axis of the anion molecules. The magnetic behavior of these salts is strongly dependent on the structural subtype, as described below.

3.4.1 $[M(Cp^*)_2][Ni(dmio)_2] \cdot$ Solv, with M = Fe, Mn and Solv = MeCN, THF, Me_2CO

The crystal structure of $[Fe(Cp^*)_2][Ni(dmio)_2]MeCN$ consists of layers composed of parallel $\cdot\cdot D^+D^+A^-D^+D^+A^-\cdot\cdot$ chains (type IVa), where side-by-side pairs of cations alternate with an anion, with a net charge (+) per repeat unit $(D^+D^+A^-)$

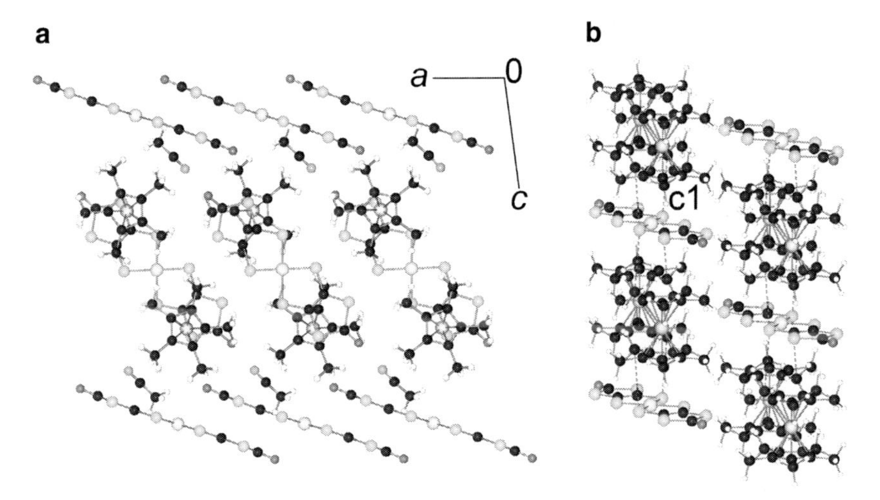

Fig. 23 (**a**) View of the crystal structure of $[Fe(Cp^*)_2][Ni(dmio)_2]MeCN$ along the stacking direction. (**b**) View of one IVa chain, c1 correspond to the closest intrachain contacts (D^+A^-)

[57]. These charged layers are separated by anion layers, as represented in the view along [0 1 0] in Fig. 23a, for $[Fe(Cp^*)_2][Ni(dmio)_2]MeCN$. The chains are regular and the Cp fragments of the cation sit above the dmio ligands of the anion. In this compound, unlike the previous compounds, the C_5Me_5 ligands from $[Fe(Cp^*)_2]^+$ present an eclipsed conformation. No interionic short contacts are observed in $[Fe(Cp^*)_2][Ni(dmio)_2]MeCN$. The closest interionic intrachain (D^+A^-) separation involves a S atom from the central MS_4 fragment of the anion and one of the C atoms from the Cp ring of the cation (c1), with a S-C distance of 3.945 Å, exceeding d_W by 15%, as shown in Fig. 23b. The chains in the layers are quite isolated and the solvent molecules are located in cavities between the $D^+D^+A^-$ chains layers and the anionic layers. In the anionic layers short A^-A^- separations involving S atoms from the five membered ring C_2S_2C of the dmio ligands, with a S-S distance of 3.777 Å, exceeding d_W by 2%. Also relatively close A^-A^- contacts between the D^+ D^+ A^- chains and the anionic layers were observed in $[Fe(Cp^*)_2][Ni(dmio)_2]$, involving O atoms from the anion layers and S atoms from the C_2S_2C ring of the dmio ligands, with a O-S distance of 3.398 Å, exceeding d_W by 5%. Similar supramolecular arrangements were observed in case of the salts $[Mn(Cp^*)_2][Ni(dmio)_2] \cdot MeCN$ [52], $[Fe(Cp^*)_2][Ni(dmio)_2] \cdot THF$ [59] and $[Mn(Cp^*)_2][Ni(dmio)_2] \cdot Me_2CO$ [54].

The magnetic behavior of the salts $[Fe(Cp^*)_2][Ni(dmio)_2] \cdot MeCN$ [57], $[Mn(Cp^*)_2][Ni(dmio)_2] \cdot MeCN$ [52, 53], $[Fe(Cp^*)_2][Ni(dmio)_2] \cdot THF$ [58, 59] and $[Mn(Cp^*)_2][Ni(dmio)_2] \cdot Me_2CO$ [54] is dominated by FM interactions, and the magnetic susceptibility of these compounds follow a Curie-Weiss behavior with θ values of 2.0, 2.8, 10.9 and 8.5 K respectively. The dominant FM interactions are ascribed to the intrachain interactions between the anions and the cations [52–54, 58, 59], in the framework of the McConnell I mechanism. In spite of the presence of shorter interionic A^-A^- contacts between the anions from the chains and within the anionic layers, these contacts must give rise to much weaker interactions than those from the intrachain D^+A^- π–π contacts. A detailed discussion of the crystal structure, namely concerning the effect of the solvent molecules in the intermolecular contacts, showed that the solvent molecules play a important role in the intermolecular magnetic coupling, which reflects in the observed θ values of these compounds [54].

3.4.2 [Fe(Cp*)$_2$][Ni(bds)$_2$]MeCN

The crystal structure of [Fe(Cp*)$_2$][Ni(bds)$_2$], as the salts described in Sect. 3.4.1, consists of 2D layers composed of parallel type IVa chains, which are separated by anion layers [15]. Besides contacts involving H atoms, no other short contacts were found in the structure. Within the IVa chains, one of the Cp fragments of the cation sits above a C_6 ring of the ligand, while the second one is displaced towards the center of the anion. For the first Cp the closest D^+A^- interatomic separation (C–C) exceeds d_W by 8%, while for the second Cp the closest D^+A^- contact (C–Se) exceeds d_W by 4%. The $D^+D^+A^-$ chains are relatively isolated and the solvent is located on cavities between the chains. No close interionic interlayer distances were observed involving molecules in the $D^+D^+A^-$ layers and the anion layers or between the anions in the anionic layers.

At high temperatures (T>25 K), χT is nearly temperature independent, but a clear increase upon cooling is observed at low temperatures, indicating that FM interactions are dominant in [Fe(Cp*)$_2$][Ni(bds)$_2$] [15]. These interactions are attributed to the D^+A^- intrachain contacts.

3.4.3 [Fe(Cp*)$_2$]$_2$[Cu(dcdmp)$_2$]

The crystal structure of [Fe(Cp*)$_2$]$_2$[Cu(dcdmp)$_2$] consists of a parallel arrangement of type IVa chains [63]. Unlike the other compounds based on type of chains, (IV) chains, with the $D^+D^+A^{2-}$ repeat unit, for this salt the chains are neutral, and there are no anionic layers. Within the chains, the cations sit on top of the pirazine rings from the ligands of the anionic complexes, with a separation between the Cp and the pirazine rings of 3.576 Å. Besides some short contacts involving H atoms, no other short contacts were observed in this compound and the chains are relatively isolated.

In the case of [Fe(Cp*)$_2$]$_2$[Cu(dcdmp)$_2$], χT is nearly temperature independent, decreasing slightly at low temperatures ($T < 30$ K). This suggests that the magnetic behavior is due to weak AF D^+A^- interactions. The isothermal obtained at 1.7 K, presents magnetization values slightly below those predicted by the Brillouin function, confirming the presence of very weak AF interactions [63].

3.4.4 $[M(Cp^*)_2][M'(bdt)_2]$ (M = Fe, Mn and Cr; M′ = Co, Ni and Pt)

The compounds $[M(Cp^*)_2][M'(bdt)_2]$ (M = Fe, Mn, and Cr; M′ = Ni and Co) exhibit related crystal structures, consisting of arrangements of layers composed by parallel type IVb chains [55, 61] separated by anionic sheets. In these IVb chains, one of the cations from the pairs is aligned with the stacking axis, while the other is nearly perpendicular (~85°) and lies along the long axis of the anions. This arrangement leads to significant distortions in the arrangement of the molecules in the chains and the angle between the average calculated planes of two successive anions is ca. 11.5°. A view along the chain direction, [1 −1 0], is shown in Fig. 24a for $[Fe(Cp^*)_2][Pt(bdt)_2]$. Disregarding contacts involving H atoms, the only short contact in the crystal structure corresponds to a Pt–C contact, involving a terminal C atom from the anion from a chain in one of the layers with a Pt atom from an anion of the anionic layer (c4). As shown in Fig. 24b, a variety of relatively close D^+A^- intrachain separations, involving S atoms from anion and C atoms from the Cp rings (c1, c1′, c2 and c2′), exceeding d_W by ~8–10%. Relatively close D^+A^- separations (c3 and c3′) were also found between the C_6 and Cp π systems from the anions (in the anionic layers) and the rotated cations (within the chains), exceeding d_W by ~8%. Within the layers, the chains are relatively isolated, but short distances between the chains and anions on the anionic layers were observed, such as the A^-A^- Pt–C contact c4; however the chain layer on top in Fig. 24a is slightly more isolated from the anionic layers, and in these the closer A^-A^- contacts correspond to the S…C contacts c4′ that exceed d_W by ~12%. In the case of $[Cr(Cp^*)_2][M'(bdt)_2]$ (M′ = Ni and Co), the crystal structure is more symmetric and there is only one type of mixed IVb chains in the structure. In all others salts (including $[Fe(Cp^*)_2][Pt(bdt)_2]$) the unit cell shows a duplication along *b*, but

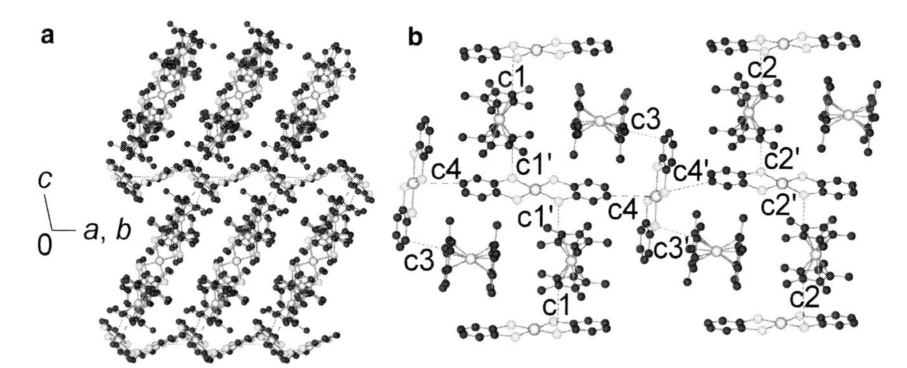

Fig. 24 **(a)** View of the crystal structure of $[Fe(Cp^*)_2][Pt(bdt)_2]$ along the chains direction (hydrogen atoms were omitted for clarity). **(b)** View of the structure showing the arrangement of two chains in neighboring chain layers and anions from the anionic layers. A variety of distinct intrachain D^+A^- contacts were observed (c1, c1′, c2 and c2′) with a good overlap between the π systems of the cations and anions. The cations perpendicular to the stacking axis also show similar contacts with anions from the anionic layers (c3 and c3′). Short A^-A^- contacts were detected between the anions in one of the IVb chain layers and anions in the anionic layers (c4), while those contacts are longer in the other chain layer (c4′)

present a similar supramolecular packing [55, 61], although in some compounds the intermolecular contacts show significant differences relative to the described above for $[Fe(Cp^*)_2][Pt(bdt)_2]$. It was not possible to determine the structure of $[Fe(Cp)_2]$ $[Co(bdt)_2]$.

In the case of the compounds based on the $[Co(bdt)_2]^-$ $S = 1$ anion, χT shows a pronounced drop below temperatures of the order of 40–50 K, which is typical of metal complexes with $S \geq 1$ exhibiting zero-field splitting [74]. Curie–Weiss fits to the magnetic susceptibility at high temperatures ($T > 150$ K) provided reasonable results with θ values of -21.6, -8.8 and -7.7 K, for $[M(Cp^*)_2][Co(bdt)_2]$, with M = Fe, Mn and Cr respectively [55, 61]. The magnetic susceptibilities of the salts $[Cr(Cp^*)_2][M'(bdt)_2]$ (M′ = Ni and Pt), follow a Curie–Weiss behavior, with θ values of 6.2 and 6.0 K [55, 61]. In the case of the compounds $[Fe(Cp^*)_2][M'(bdt)_2]$ and $[Mn(Cp^*)_2][M'(bdt)_2]$, with M′ = Ni and Pt, a minimum in the temperature dependence of χT is observed at 40, 125, 20 and 130 K for M/M′ = Fe/Ni, Fe/Pt, Mn/Ni and Mn/Pt respectively [55, 61]. For these compounds the minima are attributed to ferrimagnetic behaviors, as in $[Mn(Cp^*)_2][Pt(bdt)_2]$ a FIM ordering was observed at 2.7 K [55, 61]. The temperature dependence of a.c. susceptibility for $[Mn(Cp^*)_2][Pt(bdt)_2]$ revealed the presence of peaks both in real and imaginary components, $\chi'(T)$ and $\chi''(T)$, which is good agreement with the presence of a spontaneous magnetization in this compound, as expected for a FIM. This was also confirmed by the observation of a peak in the second harmonic of the a.c. susceptibility [55, 61]. The magnetization field dependence, at 1.8 K, is shown in Fig. 25 (squares). After a fast increase at low fields, the magnetization attains an almost

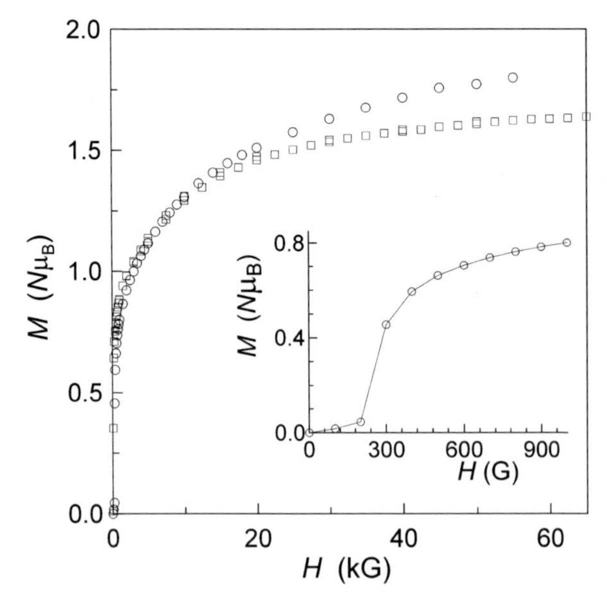

Fig. 25 Magnetization isothermal of $[Mn(Cp^*)_2][Pt(bdt)_2]$ (*squares*) and $[Fe(Cp^*)_2][Ni(bdt)_2]$ (*circles*) at 1.8 K. The *inset* shows the low field sigmoidal behavior observed in case of $[Fe(Cp^*)_2]$ $[Ni(bdt)_2]$

constant value that is in good agreement with that predicted for a FIM ordering, $M_{sat} = S_D g_D - S_A g_A \approx 1.57\ N\mu_B$, calculated for $S_A = 1/2$, $g_A = 2.06$ [75], $S_D = 1$ and $g_D = 2.6$ (due to the high anisotropy of the g value of the cation, this g_D value was obtained from susceptibility temperature dependence at high temperatures). The magnetization field dependence, at 2 K, of $[Mn(Cp^*)_2][Ni(bdt)_2]$ is also shown in Fig. 25 (circles). A metamagnetic transition was observed to occur in this compound, with $T_N = 2.3$ K and $H_C = 200$ G at 2 K [55, 61, 62], as shown in the inset of Fig. 25. The magnetization values above the critical field, in the high field state, are of the same order as the ones observed in case of $[Mn(Cp^*)_2][Pt(bdt)_2]$ and are considerably smaller than the FM saturation magnetization value, $M_{sat} = S_D g_D + S_A g_A \approx 3.53\ N\mu_B$, calculated for $S_A = 1/2$, $g_A = 2.06$ [75], $S_D = 1$ and $g_D = 2.5$ (this g_D value was obtained from the susceptibility temperature dependence at high temperature). Therefore the high field state of $[Mn(Cp^*)_2][Ni(bdt)_2]$ is consistent with a FIM state.

The complexity of the crystal structure of the $[M(Cp^*)_2][M'(bdt)_2]$ salts and the large number of intermolecular contacts in this series of complexes prevents a clear interpretation of the magnetic behavior and a correlation between the crystal structures and the magnetic properties. However, the saturation magnetization value in the case of the compounds $[Mn(Cp^*)_2][M'(bdt)_2]$ (M′ = Ni and Pt) is consistent with a D^+A^- AF coupling. A similar coupling is expected to occur in $[Fe(Cp^*)_2][M'(bdt)_2]$ (M′= Ni and Pt), where χT shows a similar behavior, typical of a FIM above the ordering temperature.

3.4.5 $[Fe(Cp^*)_2]_2[Cu(mnt)_2]$

$[Fe(Cp^*)_2]_2[Cu(mnt)_2]$ presents a crystal structure, based on an arrangement of type IVc chains. As in $[Fe(Cp^*)_2]_2[Cu(dcdmp)_2]$, the chains are neutral since the anion charge is 2^- [50]. A view along the chains direction, [1 0 1], is shown in Fig. 26a for

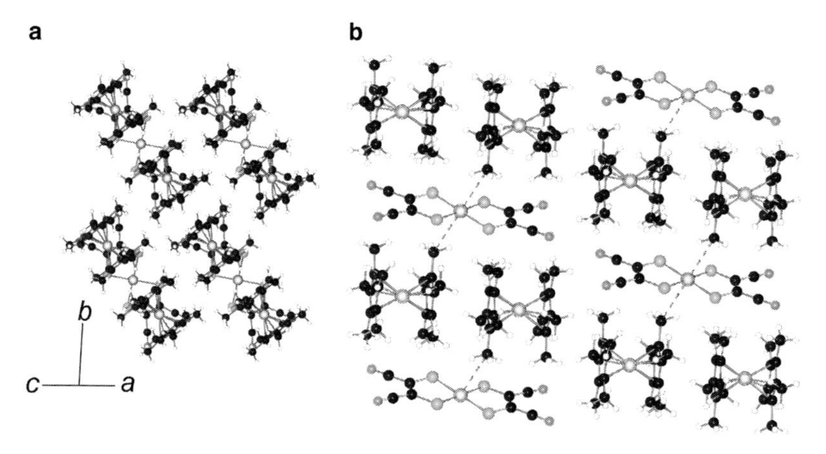

Fig. 26 (**a**) View of the crystal structure of $[Fe(Cp^*)_2]_2[Cu(mnt)_2]$ along the chain direction. (**b**) View of two type IVc neighboring chains

$[Fe(Cp^*)_2]_2[Cu(mnt)_2]$. In this compound both cations from the D^+D^+ pair in the repeat unit are perpendicular to the chain direction as shown in Fig. 26b. Short intrachain contacts were observed and involve the Cu from the anion and a C from one of the Me groups in the cation, with a Cu-C distance of 3.562 Å. The side-by-side cations are relatively separated and the closer C-C separations exceeds d_W by 20%. No short interionic intrachain contacts were observed and the chains are essentially isolated.

For $[Fe(Cp^*)_2]_2[Cu(mnt)_2]$ the magnetic susceptibility follows a Curie-Weiss behavior, with $\theta = -7.95$ K [50]. The dominant AF interactions observed in this compound are consistent with the C–S (D^+A^-) or C–C (D^+ D^+) contacts, observed in the crystal structure, as the spin density of the atoms involved in those contacts is expected to have the same signal.

3.5 *Salts with Segregated Stacks and Other Structures*

Most of the salts based on decamethylmetallocenium radical cations and on planar metal bisdichalcogenate radical anions reported so far present crystal structures with mixed linear chain basic motives. The only known exception is $[Fe(Cp^*)_2][Ni(mnt)_2]$, which exhibits another type of crystal structure based on a D^+ $[A_2]^{2-}D^+$ repeat unit [28]. In the case of this compound the magnetic behavior is dominated by the intradimer antiferromagnetic interactions.

The crystal structure was not determined in case of the compounds $[M(Cp^*)_2][Ni(tcdt)_2]$, with M = Fe, Mn and Cr. The magnetic behavior of these compounds is dominated by AF interactions, as indicated by the negative θ values, –22.9, –28.5 and –20.4 K, for M = Fe, Mn and Cr respectively.

As most of the work with this type of salts was essentially motivated by the results obtained with the salts based on decamethylmetalocenium cations and polynitrile planar radical anions, the use of different metallocenium derivatives was limited to a small number of compounds. Among these only $[Fe(C_5Me_4SCMe_3)_2][M(mnt)_2]$, M = Ni and Pt, present crystal structures based on mixed linear chain motives.

In the salts based on segregated stacks of metallocenium-based cations with $[M(mnt)_2)]^-$ (M = Ni, Pd and Pt) anions, the anionic stacks are isolated from each other by the cations; in a few compounds these stacks are regular (at high temperatures) but in most cases the $[M(mnt)_2)]^-$ units form dimers. The magnetic behavior of these compounds is dominated by the AF interactions between the anion units.

The crystal structure of $[Fe(Cp)_2]_2[Ni(mnt)_2]_2[Fe(Cp)_2]$ is composed by segregated stacks of pairs of cations, $[Fe(Cp)_2]^+$, and zig-zag dimerized anions stacks, with a neutral $[Fe(Cp)_2]$ molecule laying beside each anionic dimer [65]. In the case of $[Co(Cp)_2][Ni(dmit)_2]$, the crystal structure consists of layers composed by two types of chains formed by the $[Ni(dmit)_2]^-$ anions, through short S–S contacts. Cation pairs are located between the anionic stacks [68]. In the crystal structure of $[Co(Cp)_2][Ni(dmit)_2]_3$ 2MeCN, the partially oxidized anions form layers that are composed by parallel arrangements of chains of triads, $[Ni(dmit)_2]_3^-$. The interplanar distances within a triad and between triads are almost identical [69]. The magnetic behavior of these compounds was not reported.

In the salt $[Fe(C_5Me_4SMe)_2][Ni(mnt)_2]$ the crystal structure consists of segregated chains of cations and anions. The $[Ni(mnt)_2]^-$ units form zig-zag chains of dimers, with a Ni-S contact of 3.72 Å. χT decreases very slightly with cooling, which indicates weak AF interactions [49].

The crystal structure of $[Fe(C_5H_4R)_2][Ni(mnt)_2]$ ($[Fe(C_5R)_2]^+$ = 1,1′-bis[2-(4-(methylthio)-(*E*)-ethenyl]ferrocenium) is based on segregated chains of cations and anions. The stacks of anions are isolated by the cations and consist of a packing of dimers. In this compound χT decreases cooling, which indicating the presence of dominant AF interactions [66].

The crystal structure of the compounds $[Fe(Cp)(C_5H_4CH_2NMe_3)][M(mnt)_2]$, M = Mi, Pt, is based on segregated stacks of cations and anions. The stacks of the anions are isolated by the cationic chains, and they consist of pairs of $[M(mnt)_2]^-$ in a slipped configuration packed in stacks. Within the pairs, the $[M(mnt)_2]^-$ units interact through short M-S contacts, while there are no short contacts between the dimers. The magnetic behavior of $[Fe(Cp)(C_5H_4CH_2NMe_3)][Ni(mnt)_2]$ is consistent with the presence of magnetically isolated, AF coupled $[Ni(mnt)_2]^-$ dimers. However, in $[Fe(Cp)(C_5H_4CH_2NMe_3)][Pt(mnt)_2]$ a structural phase transition occurs at 247 K. Below this temperature its structure and magnetic behavior are consistent with isolated dimers [67].

The crystal structure of the salts [1′,1″-R_2–1,1″-biferrocene]$[Ni(mnt)_2]$ (R = isopropyl or dineopentyl) consist of segregated chains of anions and cations. For the compound with R = isopropyl the chains are based on an arrangement of slightly trimerized anions, $[Ni(mnt)_2]_3^{3-}$, with close interanionic separations, while in the case of the compound with R = dineopentyl, although the structure is still based on a similar arrangement, the anionic units are quite separated from each other. In both compounds a decrease of χT with cooling is observed, indicating dominant AF interactions [70].

The crystal structure of the compounds $[Fe(R\text{-}Cp)(R'\text{-}Cp)][Ni(mnt)_2]$ (R = H, R′ = *n*-butyl, *tert*-butyl and R = R′ = diethyl, diisopropyl) consists of isolated chains of anions that are surrounded by chains of cations. In the anionic chains the $[Ni(mnt)_2]^-$ units are dimerized through S-S or Ni-S short contacts, the dimmers interacting through slightly longer (S–S or Ni-S) contacts. χT decreases with cooling, which is consistent with dominant AF interactions [71].

4 Summary and Conclusions

The study of salts based on metallocenium cations and transition metal bisdichalcogenide anions was started almost 20 years ago and it experienced a renewed interest in the last few years, allowing a large number of compounds to be characterized. These studies, as summarized in this review, enabled it to be clearly shown in many compounds how the magnetic properties of these salts can be correlated with the crystal and molecular structures. Particularly by the study of isostructural or structurally related series of compounds with small variations in the cation and anion, it was possible to clearly put into evidence the role of variable magnetic moment and ion magnetic anisotropy in the magnetic properties of the molecular materials.

These salts present crystal structures essentially based on linear chain arrangements of alternating cation (D^+) and anion (A^-) stacks which can be classified into four major structural types, depending on the stacking motifs. These stacking patterns are determined by factors such as the dimensions of the metal bisdichalcogenate complexes, the tendency of the anions to associate as dimers and the degree of extension of the π system and the charge density distribution over the ligands.

The magnetic properties of these salts, which at low temperatures can present a wide range of magnetic ordering and associated phase transitions, are correlated with the type of magnetic interaction between the magnetic building blocks. In the large majority of cases the magnetic intermolecular interactions can be well described by the McConnell model using spin density calculations. In the case of simple structures this model is easily applied with success, but in the cases of compounds presenting more complex structures, namely with anions based on more extended ligands and with spin polarization effects, there is often competition between FM and AFM intermolecular contacts, and as a consequence a clear understanding of the magnetic behavior with that model is often not so easily achieved.

Acknowledgments

The authors wish to thank their coworkers for continued collaboration in this field, particularly D. Belo, S. Rabaça, R. Meira, I.C. Santos, M.T Duarte, J. Novoa, C. Rovira and R.T. Henriques. The financial support from Fundação para a Ciência e Tecnologia and MAGMANet network of excellence is gratefully acknowledged.

References

1. Robbins JL, Edelstein NM, Spencer B, Smart JC (1982) J Am Chem Soc 104:1882–1893
2. Robbins JL, Edelstein NM, Cooper SR, Smart JC (1979) J Am Chem Soc 101:3853–3857
3. Miller JS, Calabrese JC, Rommelmann H, Chittipeddi SR, Zhang JH, Reiff WM, Epstein AJ (1987) J Am Chem Soc 109:769–781
4. Kaul BB, Durfee WS, Yee GT (1999) J Am Chem Soc 121:6862–6866
5. Broderick WE, Thompson JA, Day EP, Hoffman BM (1990) Science 249:401–403
6. Miller JS, McLean RS, Vazquez C, Yee GT, Narayan KS, Epstein AJ (1991) J Mater Chem 1:479–480
7. Faulmann C, Rivière E, Dorbes S, Senocq F, Coronado E, Cassoux P (2003) Eur J Inorg Chem 2880–2888
8. Broderick WE, Thompson JA, Hoffman BM (1991) Inorg Chem 30:2958–2960

9. Miller JS, Callabrese JC, Rommelmann H, Chittipedi SR, Zhang JH, Reiff WM, Epstein AJ (1987) J Am Chem Soc 109:769
10. Chittipedi S, Cromack KR, Miller JS, Epstein AJ (1987) Phys Rev Lett 58:2695
11. Miller JS, Epstein AJ (1993) In: O'Connor CJ (ed) Research frontiers in magnetochemistry. World Science, Singapore, p 283
12. Miller JS, Epstein AJ (1996) In: Coronado E, Delhaès P, Gatteschi D, Miller JS (eds) Molecular magnetism: from molecular assemblies to the devices. Kluwer, Dordrecht, p 379
13. Yee GT, Miller JS (2005) In: Miller JS, Drillon M (eds) Magnetism: molecules to materials V, chap 7. Wiley, Weinheim
14. Millar JS, Calíbrese JC, Epstein AJ (1989) Inorg Chem 28:4230
15. Broderick WE, Thompson JA, Godfrey MR, Sabat M, Hoffman BM (1989) J Am Chem Soc 111:7656
16. Robertson N, Cronin L (2002) Coord Chem Rev 227:93
17. Karlin KD, Tiefel EI (eds) (2004) Dithiolene chemistry: synthesis, properties, and applications. In: Progress in inorganic chemistry, vol 52. Wiley, New York
18. Eisenberg R (1971) Prog Inorg Chem 12:295
19. Alvarez S, Ramon V, Hoffman R (1985) J Am Chem Soc 107:6253
20. Gama V, Henriques RT, Almeida M, Veiros L, Calhorda MJ, Meetsma A, de Boer JL (1993) Inorg Chem 32:3705
21. Gama V, Henriques RT, Bonfait G, Almeida M, Meetsma A, Van Smaalen S, de Boer JL (1992) J Am Chem Soc 114:1986
22. Gama V, Henriques RT, Bonfait G, Almeida M, Ravy S, Pouget JP, Alcácer L (1993) Mol Cryst Liq Cryst 234:171–178
23. Almeida M, Henriques RT (1997) In: Nalwa HS (ed) Handbook of organic conductive molecules and polymers, vol 1. Wiley, Chichester, p 87
24. Graf D, Choi ES, Brooks JS, Henriques RT, Almeida M, Matos M (2004) Phys Rev Lett 93:076406
25. Brooks JS, Graf D, Choi ES, Almeida M, Dias JC, Henriques RT, Matos M (2006) Curr Appl Phys 6:913–918
26. Cassoux P, Valade L (1996) In: Bruce DW, O'Hare D (eds) Inorganic materials, 2nd edn. Wiley, New York, p 1
27. Cassoux P, Miller JS (1998) In: Interrante LV, Hampden-Smith MJ (eds) Chemistry of advanced materials. Wiley, Weinheim, p 19
28. Belo D, Alves H, Rabaça S, Pereira LC, Duarte MT, Gama V, Henriques RT, Almeida M, Ribera E, Rovira C, Veciana J (2001) Eur J Inorg Chem 12:3127
29. Allan ML, Coomber AT, Marsden IR, Martens JHF, Friend RH, Charlton A, Underhill AE (1993) Synth Met 55/57:3317
30. Coomber AT, Beljonne D, Friend RH, Bredas JL, Charlton A, Robertson N, Underhill AE, Kurmoo M, Day P (1996) Nature 380:144–146
31. Xie J, Ren X, Song Y, Zou Y, Meng Q (2002) J Chem Soc Dalton Trans 2868
32. Fujiwara E, Yamamoto K, Shimamura M, Zhou B, Kobayashi A, Takahashi K, Okano Y, Cui H, Kobayashi H (2007) Chem Mat 19:553–558
33. Nunes JPM, Figueira MJ, Belo D, Santos IC, Ribeiro B, Lopes EB, Henriques RT, Vidal-Gancedo J, Veciana J, Rovira C, Almeida M (2007) Chem Eur J:9841–9849
34. Gama V, Duarte MT (2005) In: Miller, Drillon M (eds) Magnetism: molecules to materials V, chap 1. Wiley, Weinheim
35. McConnell HM (1963) J Chem Phys 39:1910
36. McConnell HM, Proc Robert A (1967) Welch Found Conf Chem Res 11:144
37. Miller JS, Epstein AJ (1994) Angew Chem Int Ed Engl 33:385
38. Kahn O (1993) Molecular magnetism, chap 12. VCH, New York
39. Deumal M, Novoa JJ, Bearpark MJ, Celani P, Olivucci M, Robb MA (1998) J Phys Chem A 102:8404
40. Deumal M, Cirujeda J, Veciana J, Novoa JJ (1999) Chem Eur J 5:1631

41. Gama V, Belo D, Rabaça S, Santos IC, Alves H, Duarte MT, Waerenborgh JC, Henriques RT (2000) Eur J Inorg Chem 9:2101
42. Broderick WE, Thompson JA, Hoffman BH (1991) Inorg Chem 30:2960
43. Miller JS, private communication
44. Rabaça S, Meira R, Pereira LCJ, Duarte MT, Novoa JJ, Gama V (2001) Inorg Chim Acta 326:89
45. Rabaça S, Vieira BJC, Meira R, Santos IC, Pereira LCJ, Duarte MT, Gama V (2008) Eur J Inorg Chem 3839–3851
46. Gama V, Rabaça S, Ramos C, Belo D, Santos IC, Duarte MT (1999) Mol Cryst Liq Cryst 335:81–90
47. Belo D, Mendonça J, Santos IC, Pereira LCJ, Almeida M, Novoa JJ, Rovira C, Veciana J, Gama V (2008) Eur J Inorg Chem 5327–5337
48. Belo D (2001) Ph.D. Dissertation, Technical University of Lisbon, Portugal
49. Zürcher S, Gramlich V, Arx D, Togni A (1998) Inorg Chem 37:4015
50. Fettouhi M, Ouahab L, Hagiwara M, Codjovi E, Kahn O, Constat-Machado H, Varret F (1995) Inorg Chem 34:4152
51. Rabaça S, Gama V, Belo D, Santos IC, Duarte MT (1999) Synh Met 103:2303
52. Faulmann C, Rivière E, Dorbes S, Senocq F, Coronado E, Cassoux P (2003) Eur J Inorg Chem:2880–2888
53. Faulmann C, Pullen AE, Rivière E, Journaux Y, Retailleau L, Cassoux P (1999) Synth Met 103:2296–2297
54. Faulmann C, Dorbes S, Rivière E, Andase A, Cassoux P, Valade L (2006) Inorg Chim Acta 359:4317–4325
55. Rabaça S (2003) Ph.D. Dissertation, Technical University of Lisbon, Portugal
56. Rabaça S, Santos IC, Duarte MT, Gama V, unpublished
57. Fettouhi M, Ouahab L, Codjovi E, Kahan O (1995) Mol Cryst Liq Cryst 273:29
58. Rabaça S, Santos IC, Duarte MT, Gama V (2003) Synth Met 135–236:695–696
59. Rabaça S, Santos IC, Duarte MT, Gama V (2006) Acta Crystallogr C 62:M278–M280
60. Rabaça S, Santos IC, Duarte MT, Gama V (2007) Inorg Chim Acta 360:3855–3860
61. Rabaça S, Belo D, Alves H, Pereira LCJ, Santos IC, Duarte MT, Gama V, unpublished
62. Gama V, Belo D, Santos IC, Henriques RT (1997) Mol Cryst Liq Cryst 306:17–24
63. Belo D, Figueira MJ, Santos IC, Gama V, Pereira LC, Henriques RT, Almeida M (2005) Polyhedron 24:2035–2042
64. Sano M, Adachi H, Yamatera H (1981) Bull Chem Soc Jpn 54:2636
65. Day MW, Qin J, Yang C (1998) Acta Crystallogr C 54:1413
66. Hobi M, Zürcher S, Gramlich V, Burckhardt U, Mensing C, Spahr M, Togni A (1996) Organometallics 15:5342
67. Pullen A, Faulmann C, Pokhodnya KI, Cassoux P, Tokumoto M (1998) Inorg Chem 37:6714
68. Qi F, Xiao-Zeng Y, Jin-Hua C, Mei-Yun H (1993) Acta Crystallogr C 49:1347
69. Faulmann C, Delpech F, Malfant I, Cassoux P (1996) J Chem Soc Dalton Trans:2261
70. Mochida T, Takazawa K, Matsui H, Takahashi M, Takeda M, Sato M, Nishio Y, Kajita K, Mori H (2005) Inorg Chem 44:8628–8641
71. Mochida T, Koinuma T, Akasaka T, Sato M, Nishio Y, Kajita K, Mori H (2007) Chem Eur J 13:1872–1881
72. Stryjewski E, Giordano N (1977) Adv Phys 26:487–650
73. Belo D, Pereira LCJ, Almeida M, Rovira C, Veciana J, Gama V (2009) Dalton Transactions (in press)
74. van der Put PJ, Schilperoord AA (1974) Inorg Chem 13:2476–2481
75. McCleverty JA (1968) In: Cotton FA (ed) Progress in inorganic chemistry, vol 10. Interscience, New York, p 49

Top Organomet Chem (2009) 27: 141–159

Conductive Materials Based on $M(dmit)_2$ Complexes and Their Combination with Magnetic Complexes

Lydie Valade and Christophe Faulmann

Abstract Metal complexes including the $dmit^{2-}$ (1,3-dithio-2-thione-4,5-dithiolato) ligand are the only class of metal bis-dithiolenes to give rise to superconductive molecular materials. This chapter first focuses on the description of these superconductive phases. Further sections describe the association of $M(dmit)_2$ moieties with three types of magnetic molecules, i.e., metalloceniums, radical cations, and spin crossover complexes.

Keywords $M(dmit)_2$, Metal bis(dithiolene), Metallocenium, Radical cations, Spin crossover, Magnetic properties, Superconductor

Contents

1 Introduction

The first molecular conductor based on a metal complex was prepared by Knop [1]. By oxidizing $K_2[Pt(CN)_4]$ (KCP) with chlorine or bromine, he observed the

L. Valade and C. Faulmann(✉)
CNRS; LCC (Laboratoire de Chimie de Coordination); 205 route de Narbonne, F-31077 Toulouse, France Université de Toulouse; UPS, INPT; LCC; F-31077 Toulouse, France,
E-mail: lydie.valade@lcc-toulouse.fr
Laboratoire de Chimie de Coordination, CNRS-UPR8241, 205 route de Narbonne, 31077 Toulouse Cedex 4, France,
E-mail: valade@lcc-toulouse.fr

M. Fourmigué and L. Ouahab (eds.), *Conducting and Magnetic Organometallic Molecular Materials*, Topics in Organometallic Chemistry 27,
DOI: 10.1007/978-3-642-00408-7_6,

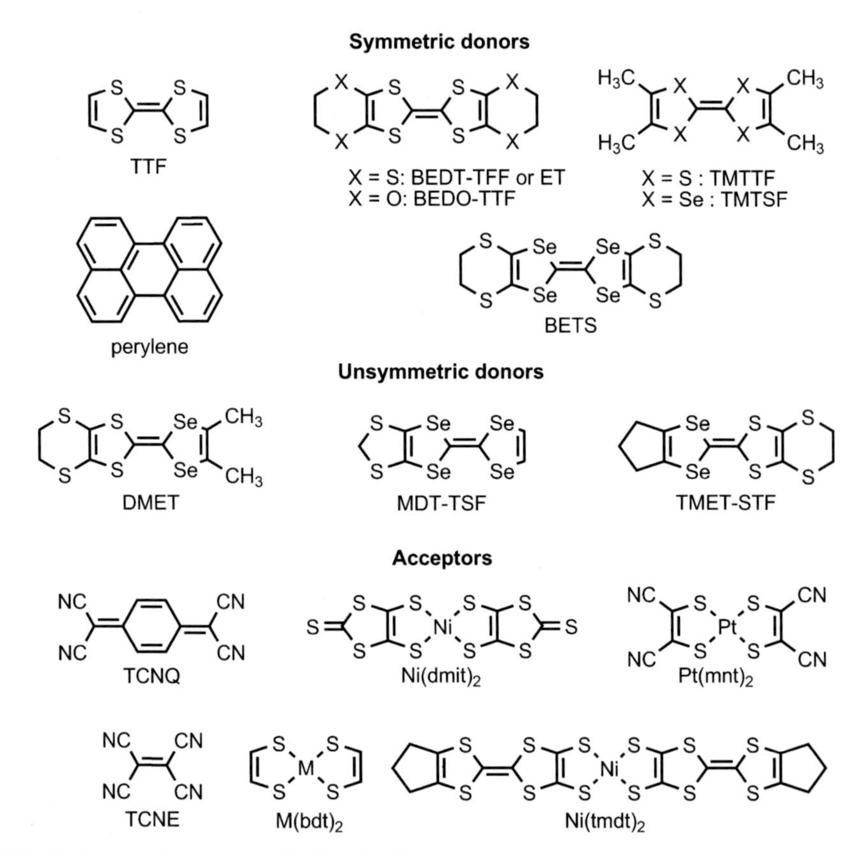

Fig. 1 Donor and acceptor molecules cited in text

formation of copper-shining crystals. It took 130 years before Zeller [2, 3] measured the conductivity of these crystals and evidenced their metallic behavior in the high temperature domain. In KCP crystals, $[Pt(CN)_4]$ complexes form stacks along which electron delocalization proceeds through the overlap of d_{z^2} orbitals. The material is essentially one-dimensional: the room temperature conductivity along the stacks is 300 S cm^{-1} and 10^{-5} lower in directions perpendicular to the stacks. None of the KCP derivatives retain its metallic behavior at low temperatures. Because of lattice distortion, a metal-to-insulator transition occurs, a feature predicted by Peierls [4], and inherent to one-dimensional (1D) systems. The first organic metal, TTF-TCNQ (Fig. 1), isolated in 1973 [5, 6], also undergoes the so-called Peierls distortion which occurs through the successive blocking of the donor TTF and acceptor TCNQ stacks. Rapidly, the selection of molecular building blocks for preparing metallic molecular systems stable at low temperatures was oriented towards molecules able to lead to 2D or 3D electronic networks. Se analogs of the TTF donor and an increase of the number of S atoms in the parent molecules were undertaken. A comprehensive review of all modifications conducted on the TTF core can be found in a book edited by Yamada and Sugimoto [7].

Superconductivity was reached by using many donors such as, TMTSF, BEDT-TTF, BETS, DMET, BEDO-TTF, MDT-TSF, DMET-STF, . . .[8]. The first molecular superconductor at ambient pressure $(TMTSF)_2ClO_4$, was isolated by Bechgaard [9]. Although an intermolecular 2D network involving peripheral chalcogen atoms appears between stacks, Bechgaard salts are essentially one-dimensional. Two-dimensional systems really appeared with the BEDT-TTF family of superconductors and later on in the BETS family [10]. A comprehensive review has been published recently by Saito and Yushida [8].

Metal dithiolenes complexes have been studied for a long time. However, they entered the field of molecular materials only at the end of the 1960s when their electrical and magnetic properties were studied. Their properties have been reviewed recently [11–14]. The first dithiolene complex exhibiting high conductivity (700 S cm^{-1}) and a metallic behavior down to 8.2 K is $(Per)_2[Pt(mnt)_2]$, prepared in 1980 [15]. These complexes also appear in interesting magnetic systems such as spin-Peierls, spin-ladder, and ferromagnetic systems. The $M(bdt)_2$ core of metal bis-dithiolenes is isolobal with the TTF molecule. Moreover, the presence of peripheral chalcogen atoms in metal bis-dithiolenes was thought to favor the build up of multidimensional electronic networks. Following this hypothesis, the dithiolene family based on the dmit ligand was selected at the beginning of the 1980s. The $M(dmit)_2$ complexes, first prepared by Steimecke [16], bear ten peripheral S atoms. Many fractional oxidation state complexes and donor acceptor compounds were prepared from these complexes. The dmit family is the only dithiolene family that gave rise to superconductive phases. These phases have been reviewed in [13, 17, 18]. The first one, $TTF[Ni(dmit)_2]_2$, was isolated in 1986 [19, 20]. Main properties of $M(dmit)_2$ based superconductors will be summarized in Sect. 2 of this chapter.

After the search for single properties in molecular materials, increasing interest has been devoted to multiproperty systems. A few exceptions apart, as for example the first unimolecular metallic material $Ni(tmdt)_2$ [21], neutral TTF–betainic radicals [22], and complexes as $[Cu(hfac)_2(TTF\text{-}py)_2](X)_2.2CH_2Cl_2$ ($X = PF_6, BF_4$) [23, 24], and $[(ppy)Au(C_8H_4S_8)]_2[PF_6]$ [25], which include both the conductive and magnetic moieties chemically bonded within a single molecule, the majority of molecular materials is built from the association of two components. Therefore, research focused on the association of components affording two different properties. Magnetic conductors were the widely studied systems [26, 27]. Combination of magnetism and conductivity, in particular that of ferromagnetism and superconductivity, first appeared as oxymoron as the internal field created by the spin carrying molecules would destroy or avoid existence of the Cooper pairs responsible for superconductivity. In molecular materials, chains exhibiting electronic delocalization may be associated with magnetic chains or isolated magnetic ions. Interactions between chains are generally weak and both properties may coexist. The main series of magnetic conductors include salts of the donors BEDT-TTF and BETS with tris-oxalatometalate anions and halometalate anions, respectively [26]. Superconductive phases have been isolated in both of these series. Acceptor molecules such as $M(dmit)_2$ have also been associated with magnetic cations such as metalloceniums [28]. Another type of magnetic species interesting to combine with conductive chains is spin crossover complexes. The spin crossover phenomenon, a cooperative phenomenon in solids, has attracted much interest in a fundamental point of view and more recently because of its potential application in switching or sensing devices and for data storage [29]. The third section of this chapter will be devoted to compounds combining metalloceniums and $M(dmit)_2$, the fourth will concern radicals and $M(dmit)_2$, and the final section will report on the properties of spin crossover transition metal complexes associated with $M(dmit)_2$.

2 $M(dmit)_2$-Based Superconductors

$M(dmit)_2$ complexes are the only metal bis-dithiolenes known to lead to superconductive phases. Twelve compounds have been isolated up to now (Table 1).

The first superconductor of the series, $TTF[Ni(dmit)_2]_2$, is a donor-acceptor type phase and was isolated in 1986 [19, 20]. Superconductivity occurs at 1.62 K under a hydrostatic pressure of 7 kbar. The pressure-temperature phase diagram of this quasi-one-dimensional superconductor was deeply studied by a.c. resistivity measurements up to 14 kbar [40]. Numerous electronic phase transitions occur upon increasing applied pressures: from high-temperature metal to successively metallic, semimetallic, semiconductive and reentrant superconductive ground states. From this rather complicated phase diagram, two unique features have been observed: the superconductive transition temperature Tc increases with pressure and, at pressures around 5.3 kbar, the superconductive ground state is reentrant into the low-pressure insulating state (Fig. 2). Also observed in the α'-$(TTF)[Pd(dmit)_2]_2$ phase, this was explained by the occurrence of CDW instabilities connected to a unique multisheets Fermi surface based on both the HOMO and the LUMO bands of the building blocks [41].

Table 1 Superconductors based on $M(dmit)_2$ complexes

Compound	$\sigma_{RT/1bar}$ (Scm^{-1})	Tc (K)	P (kbar)	Ref.
$TTF[Ni(dmit)_2]_2$	300	1.62	7	[19]
α-$(TTF)[Pd(dmit)_2]_2$	800	1.7	22	[30]
α'-$(TTF)[Pd(dmit)_2]_2$	600	5.93	24	[31]
α-(EDT–TTF)$[Ni(dmit)_2]$	100	1.3	Amb.	[32]
α-$(Et_2Me_2N)[Pd(dmit)_2]_2$	10–80	4	2.4	[33]
β'-$(Et_2Me_2P)[Pd(dmit)_2]_2$	10	4	6.9	[34]
$(EtMe_3P)[Pd(dmit)_2]_2$	13	5	3.3	[35]
β'-$(EtMe_3As)[Pd(dmit)_2]_2$	13	4.3	7	[35]
$(Me_4N)[Ni(dmit)_2]_2$	60	5	7	[36, 37]
β-$(Me_4N)[Pd(dmit)_2]_2$	30	6.2	6.5	[38]
β'-$(Me_4As)[Pd(dmit)_2]_2$	10	4	7 (uniaxial)	[39]
β'-$(Me_4Sb)[Pd(dmit)_2]_2$	10	3	10	[34]

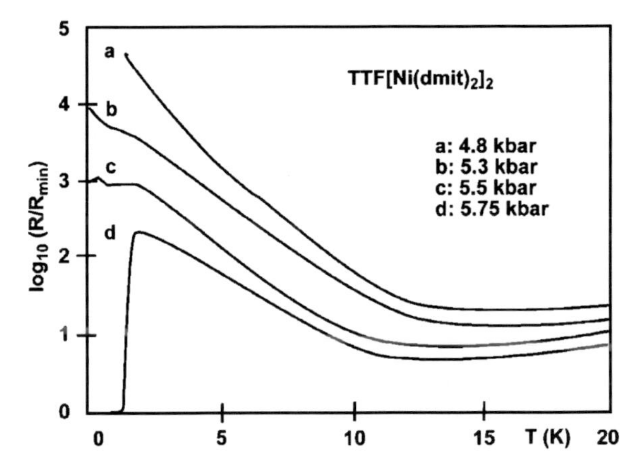

Fig. 2 Pressure-induced superconductivity in TTF[Ni(dmit)$_2$]$_2$. Redrawn after [40]

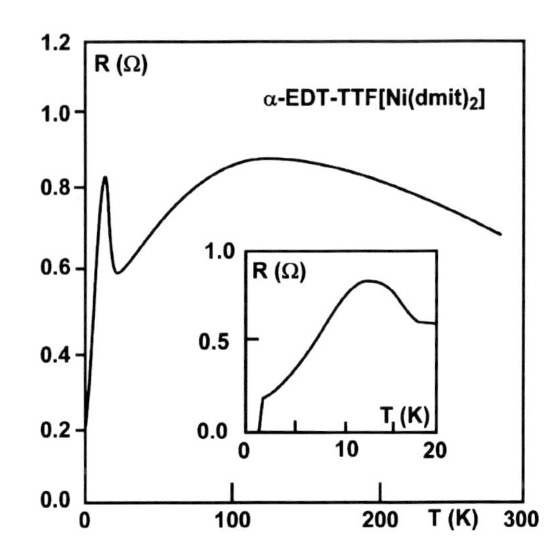

Fig. 3 Ambient-pressure superconductivity in α-(EDT−TTF)[Ni(dmit)$_2$]. The *inset* shows the superconductive transition at 1.3 K. Redrawn after [42]

Among the M(dmit)$_2$-based superconductors, α-(EDT-TTF)[Ni(dmit)$_2$] is also of donor-acceptor type and has two outstanding features; it is the only one to contain a 1:1 molar ratio of donor and acceptor units and to exhibit superconductivity at ambient pressure [32]. It was found to be superconductive below 1.3 K under ambient pressure (Fig. 3).

α-(EDT-TTF)[Ni(dmit)$_2$] exhibits a unique metallic behavior with a characteristic resistivity peak at around 14 K. Magnetoresistance studies evidenced that the conduction mainly takes place along the Ni(dmit)$_2$ stacks below 10 K while both Ni (dmit)$_2$ and EDT-TTF stacks are involved above 20 K [42, 43].

Pd(dmit)$_2$ complexes afforded the larger number of superconductive phases, the majority with closed-shell cations [13]. The later isolated phase, (EtMe$_3$P)[Pd (dmit)$_2$]$_2$, becomes superconductive at 5 K under 3.3 kbar [35]. This phase exhibits

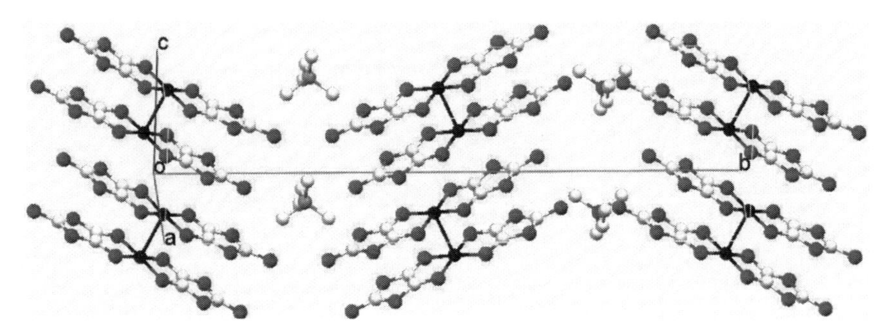

Fig. 4 Crystal structure of $(EtMe_3P)[Pd(dmit)_2]_2$ viewed from the *a**-axis. After [35]

an interesting packing organization showing uniformly stacked dimers (Fig. 4). Intra- and interdimer HOMO-HOMO overlap integrals indicate that $Pd(dmit)_2$ molecules are strongly dimerized. Therefore, each dimer can be considered as the building unit. As a matter of fact, the crystal structure of the phase performed at 10 K, reveals pairing of dimers.

3 M(dmit)$_2$ and Metalloceniums

In this section, we will focus only on compounds based on $M(dmit)_2$ anions and $(Cp^*{}_2Mn)^+$, since Gama and Almeida also report on metal-dithiolene containing compounds with metalloceniums in Chapter (see pp. 97–140).

Following the idea of McConnell [44] who proposed in 1967 that a stacking arrangement such as $\ldots D^+ A^- D^+ A^- D^+ A^- \ldots$ of donor and acceptor could lead to molecule-based magnets, the combination between metal bis-dithiolene complexes and metalloceniums was performed originally only to promote cooperative ferromagnetic interactions [45].

Using this concept, numerous compounds with formula $(Cp^*{}_2M')[M(bdt)_2]$ have been obtained (*bdt* stands for ligand 1,2-bis-dithiolene) (see Table 6 in [11]). In 1990, Broderick et al. suggested [46] that the use of a large spin metallocenium such as $(Cp^*{}_2Mn)^+$ ($S = 1$) instead of $(Cp^*{}_2Fe)^+$ ($S = 1/2$) could lead to ferromagnetic compounds with a high ordering temperature. So, after the synthesis of $(Cp^*{}_2Fe)[Ni(tfd)_2]$ in 1989 [45], (tfd^{2-} = 1,2-bis(trifluoromethyl)ethylenedithiolate $[(CF_3)_2C_2S_2]^{2-}$) the same authors reported on $(Cp^*{}_2Mn)[Ni(tfd)_2]$ [47]: indeed, whereas the $(Cp^*{}_2Fe)$-based compound only exhibits ferromagnetic interactions, the $(Cp^*{}_2Mn)$-based compound behaves like a metamagnet.

It is only 10 years after the synthesis of $(Cp^*{}_2Fe)[Ni(dmit)_2]$ [48] that the parent compound with $(Cp^*{}_2Mn)$ has been synthesized [49, 50], with the ultimate goal of combining magnetism and conductivity in the same material. As in the previous series with the tfd^{2-} ligand, $(Cp^*{}_2Mn)[Ni(dmit)_2]$ exhibits improved magnetic

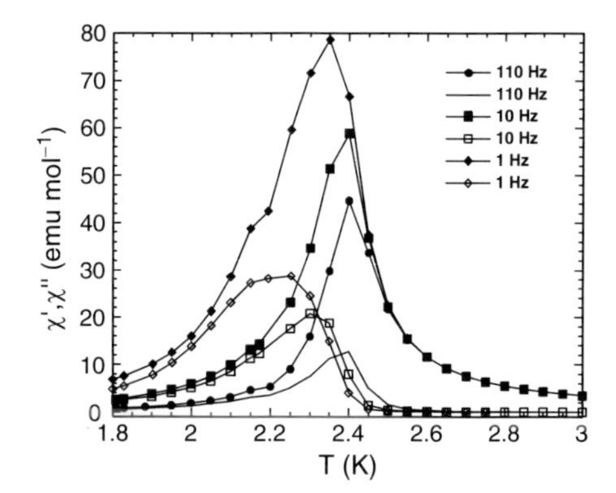

Fig. 5 AC susceptibility for $(Cp*_2Mn)[Ni(dmit)_2]$ (from [46])

properties compared to $(Cp*_2Fe)[Ni(dmit)_2]$. The latter only exhibits ferromagnetic interactions whereas $(Cp*_2Mn)[Ni(dmit)_2]$ behaves like a bulk ferrimagnet [50], as confirmed by the a.c. measurements shown in Fig. 5: an out-of-phase signal is observed at 2.5 K, indicating a ferromagnetic ordering below this temperature.

Unfortunately, whatever the metallocenium and the metal bis-dithiolene moiety, no fractional oxidation state compound with formula $(Cp*_2M')_x[M(bdt)_2]$ (with $x < 1$) has been reported in the literature. Concerning $(Cp*_2M')[Ni(dmit)_2]$ (M' = Fe, Cr, Mn), all our attempts to obtain chemically or electrochemically $(Cp*_2M')_x[Ni(dmit)_2]$ (with $x < 1$) have resulted in the synthesis of noncrystalline samples as sticky powders or mixture of fibers and grains, whose characterization was not possible. Therefore the combination between M(dmit)$_2$ and metallocenium has been unsuccessful to obtain magnetic molecular conductors.

4 M(dmit)$_2$ and Radicals

Radicals have been known for many years to form organic paramagnetic materials with numerous magnetic properties (ferro- or ferri-magnetism, spin Peierls transition, spin frustration, spin ladder systems) (see [51–60] for verdazyl radicals, [61–68] for thiazyl radicals, [69] for nitronyl nitroxide and [70–78] for Tempo radicals) (Fig. 6). When they are in their cationic form, they are valuable candidates for an association with the M(dmit)$_2$ systems: they will then provide the magnetic properties thanks to their free electron(s), whereas the M(dmit)$_2$ moieties will provide the electrical properties.

To our knowledge, there are less than 30 compounds based on radical-cations and M(dmit)$_2$ systems (Table 2). Most of them contain divalent or monovalent M(dmit)$_2$ units, and only a few of them have been structurally and magnetically characterized. Since they are not in a fractional oxidation state, they behave as insulators with low room-temperature conductivity.

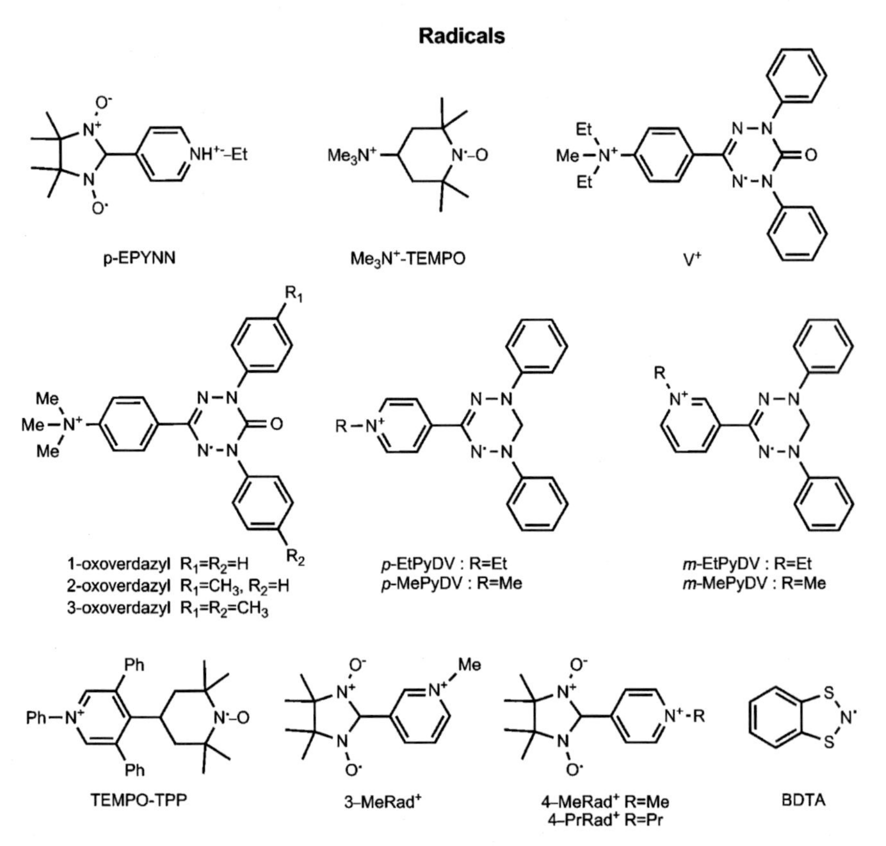

Fig. 6 Radicals cited in text

Their magnetic behaviors are dominated in most cases by AFM interactions. Only three of them, namely (p-EPYNN)[M(dmit)$_2$] (M = Ni, Au) and [4-PrRad][Ni(dmit)$_2$] exhibit FM interactions in a narrow range of temperature. The lack of structural details for [4-PrRad][Ni(dmit)$_2$] (synthesized as a powder) prevents any interpretation of its magnetic behavior.

In contrast, within (p-EPYNN)[Ni(dmit)$_2$], first synthesized in 1996 [79], it has been proven that spin-ladder chains of the Ni(dmit)$_2$ moiety coexist with the ferromagnetic one-dimensional chain of the p-EPYNN radical cation. Spin-ladders are of interest because of their potential applications in the area of quantum magnets and because it has been predicted that holes doped into even-leg ladders may pair and possibly superconduct [90–92].

Evidently, the most interesting materials are those in a fractional oxidation state, with general formula (cation)[M(dmit)$_2$]$_n$ ($n > 1$), since they can exhibit both electrical and magnetic properties. Only eight such complexes have been reported so far. All of them but (BDTA)[Ni(dmit)$_2$]$_2$ [89] have been obtained as powders. They have in general been poorly characterized, and their stoichiometries have been determined from elemental analysis. Among these powdered compounds, the

Table 2 Radical cations with M(dmit)$_2$

Compound	σ_{RT}/S cm^{-1}	Electrical behavior	Magnetic interactions	Ref.
(*p*-EPYNN)[Ni(dmit)$_2$]	4.0×10^{-7} to 1.3×10^{-4}	SC	AFM (300–150 K) FM (40–4 K)	[79, 80]
(*p*-EPYNN)[Au(dmit)$_2$]	–	–	AFM (300–40 K) FM (40–4 K)	[80]
(Me$_3$N-TEMPO)$_2$[Ni(dmit)$_2$]	–	–	–	[81–83]
(Me$_3$N-TEMPO)$_2$[Pd(dmit)$_2$]	–	–	–	[81–83]
(Me$_3$N-TEMPO)[Ni(dmit)$_2$]	4.0×10^{-3}	SC	–	[81–83]
(Me$_3$N-TEMPO)[Ni(dmit)$_2$]$_2$	1.0×10^{-2} (p)	–	Weak FM	[81–83]
(Me$_3$N-TEMPO)[Pd(dmit)$_2$]$_4$	1.0×10^{-2} (p)	–	Weak FM	[81–83]
[V][Ni(dmit)$_2$]	3.6×10^{-7}	Insulator	AFM	[84]
[V][Ni(dmit)$_2$]$_3$	8.9×10^{-2} (p)	SC	AFM	[84]
[V]$_2$[Zn(dmit)$_2$]	3.0×10^{-4}	Insulator	AFM	[84]
[V]$_2$[Pd(dmit)$_2$]	1.3×10^{-4}	SC	AFM	[84]
[V]$_2$[Pt(dmit)$_2$]	1.8×10^{-7}	Insulator	AFM	[84]
(1 Oxoverdazyl)[Ni(dmit)$_2$]	1.9×10^{-7}	Insulator	AFM	[85]
(2 Oxoverdazyl)[Ni(dmit)$_2$]	1.8×10^{-7}	Insulator	AFM	[85]
(3 Oxoverdazyl)[Ni(dmit)$_2$]	4.7×10^{-5}	Insulator	AFM	[85]
(1 Oxoverdazyl)[Ni(dmit)$_2$]$_3$	0.10 (p)	SC	AFM	[85]
[*p*-EtPyDV][Ni(dmit)$_2$]	2.3×10^{-5}	SC	AFM	[86]
[*m*-EtPyDV][Ni(dmit)$_2$]	1.1×10^{-3}	SC	AFM	[86]
[*p*-MePyDV][Ni(dmit)$_2$]	1.4×10^{-5}	SC	AFM	[86]
[*m*-MePyDV][Ni(dmit)$_2$]	5.4×10^{-6}	SC	AFM	[86]
[*p*-MePyDV][Ni(dmit)$_2$]$_3$	1.0 (p)	SC	AFM	[86]
(Tempo-TPP)[Ni(dmit)$_2$] (CH$_2$Cl$_2$)$_{0.75}$	–	–	AFM	[87]
(Tempo-TPP)[Ni(dmit)$_2$]$_6$	1.5 (p)	SC	–	[87]
(Tempo-TPP)$_2$[Pd(dmit)$_2$]	–	–	AFM	[87]
(Tempo-TPP)[Pd(dmit)$_2$]$_6$	6.44 (p)	SC	–	[87]
[3-MeRad][Ni(dmit)$_2$]	10^{-3}	SC	AFM	[88]
[4-MeRad][Ni(dmit)$_2$]	10^{-5}	SC	AFM	[88]
[4-PrRad][Ni(dmit)$_2$]	10^{-3}	SC	FM (300–40 K) AFM (40–4 K)	[88]
(BDTA)[Ni(dmit)$_2$]$_2$	0.1	SC	AFM	[89]

(p) powder; *SC* semiconductor; *AFM* and *FM* antiferromagnetic and ferromagnetic interactions, respectively

most conductive is (Tempo-TPP)[Pd(dmit)$_2$]$_6$, which exhibits a room-temperature conductivity of 6.44 S cm^{-1} [87]. Its magnetic properties have not been reported.

In contrast, (BDTA)[Ni(dmit)$_2$]$_2$ has been fully characterized [89] (X-ray structure, magnetic susceptibility data, band structure and conductivity). It exhibits an SC behavior with a room-temperature conductivity of 0.1 S cm^{-1}. Its magnetic behavior is dominated by AFM interactions, probably due to the coupling between the Ni(dmit)$_2$ moieties. Although its properties are not spectacular, this complex is the first well-characterized example of a salt containing partially-oxidized Ni(dmit)$_2$ moieties with a radical cation.

5 M(dmit)$_2$ and Spin Crossover Complexes

Multifunctional materials based on metal complexes exhibiting switching behavior are of great interest for future generations of electronic devices. Among switchable molecule-based materials, the most spectacular examples are given by metal complexes undergoing spin crossover (SCO). Various stimuli can be applied to switch the material between low-spin (LS) and high-spin (HS) states: variation of temperature, of pressure, or light irradiation [93]. Upon LS $\leftrightarrow$ HS transition, changes in the magnetism, color and structure of the material are observed. When the molecular structural changes associated with the SCO phenomenon are efficiently transmitted through the lattice, the resulting cooperativity is manifested by a strong variation of the magnetic properties and a large hysteresis [94]. From these singular properties, applications can be forecast in the field of memories, sensors, thermal displays, etc. [95–97].

The cationic nature of SCO complexes makes them good candidates to investigate the coexistence of conductive and switching properties within a single material.

[Fe(abpt)$_2$(TCNQ)$_2$], (Fig. 7; abpt = 3,5-bis(pyridin-2-yl)-4-amino-1,2,4-triazole) synthesized in 1996 by Kunkeler et al. [98] is the first compound containing a possible conductive unit (TCNQ) and an SCO moiety (Fe^{II}(abpt)$_2$). However,

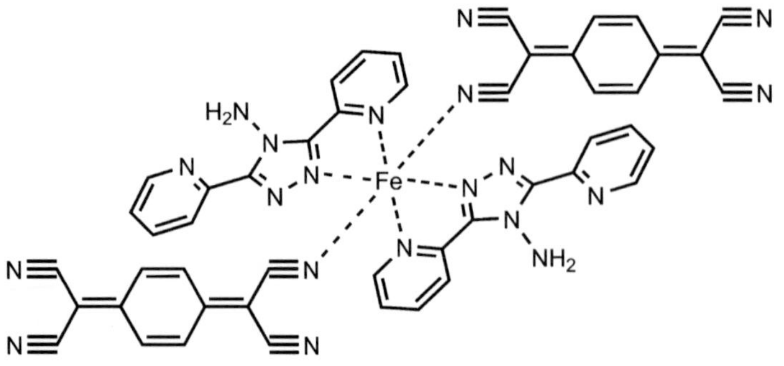

Fig. 7 [Fe(abpt)$_2$(TCNQ)$_2$]

Table 3 Spin crossover cations with $M(dmit)_2$

Compound	σ_{RT} (S cm^{-1})	Electrical behavior	Magnetic properties	Ref.
$[Fe(sal_2\text{-trien})][Ni(dmit)_2]$	–	–	Abrupt ST with a 30 K hysteresis	[99]
$[Fe(sal_2\text{-trien})][Ni(dmit)_2]_3$	0.1	SC	Paramagnet	[101]
$[Fe(salten)Mepepy][M(dmit)_2].CH_3CN$ M = Ni, Pd, Pt	–	–	ST (350–20 K)	[102]
$[Fe(salten)Mepepy][Ni(dmit)_2]_3$	0.1	–	AFM	[102]
$[Fe(qsal)_2][Ni(dmit)_2]$	–	–	ST (300–10 K)	[103]
$[Fe(3\text{-MeO-salEen})_2][Ni(dmit)_2].CH_3OH$	–	–	ST (400–200 K)	[104]
$[Fe(salEen)_2]_2[Ni(dmit)_2](NO_3).CH_3CN$	–	–	ST (400–300 K)	[104]
$[Fe(salEen)_2]_2[Ni(dmit)_2]_5.6CH_3CN$	0.12	SC	ST (350–50 K)	[104]
$[Fe(qsal)_2][Ni(dmit)_2].2CH_3CN$	–	–	ST (300–150 K), hysteresis, LIESST effect	[105]
$[Fe(qsal)_2][Ni(dmit)_2]_3.CH_3CN.H_2O$	2	SC	ST (300–100 K) LIESST effect	[106]
$[Fe(qnal)_2][Pd(dmit)_2]_3.(CH_3)_2CO$	1.6×10^{-2}	SC	ST (240–180 K)	[107]

ST spin transition; *SC* semiconductor; *AFM* antiferromagnetic interactions

although it displays a gradual spin transition between 400 and 200 K, it does not exhibit any electrical properties. Further work has never been reported in this line since, originally, $[Fe(abpt)_2(TCNQ)_2]$ was synthesized to study the possible direct magnetic exchange between coordinated TCNQ and the Fe^{II} SCO center.

To our knowledge, there are only 13 complexes based on the $M(dmit)_2$ moiety with SCO cations (Table 3).

The first metal bis-dithiolene complex with an SCO moiety as counter-ion was reported in 2005 [99], namely $[Fe(sal_2\text{-trien})][Ni(dmit)_2]$ (Fig. 8). Usually, $[Fe(sal_2\text{-trien})]^+$ is a Fe^{III} complex which exhibits very modest magnetic properties: a gradual and incomplete spin transition is observed when it is combined with $(PF_6)^-$ and $(BPh_4)^-$, and the complex remains in the LS state with Cl^-, I^- and $(NO_3)^-$.

On the other hand, when associated with $[Ni(dmit)_2]^-$, $[Fe(sal_2\text{-trien})]^+$ exhibits a cooperative spin transition behavior with a wide hysteresis loop (30 K) around 240 K (Fig. 9).

$[Fe(sal_2\text{-trien})][Ni(dmit)_2]$ is one of the rare examples of Fe^{III} complexes [100] that displays strong cooperativity and wide hysteresis. Actually, the geometrical and electronic nature of the anion $[Ni(dmit)_2]^-$ favors the occurrence of segregated stacks of anions and cations (Fig. 10). Consequently, the latter are strongly coupled via π-stacking and the large intramolecular structural modifications upon SC are transmitted quite efficiently to the whole crystal and represent the coercive force responsible for the observed large hysteresis.

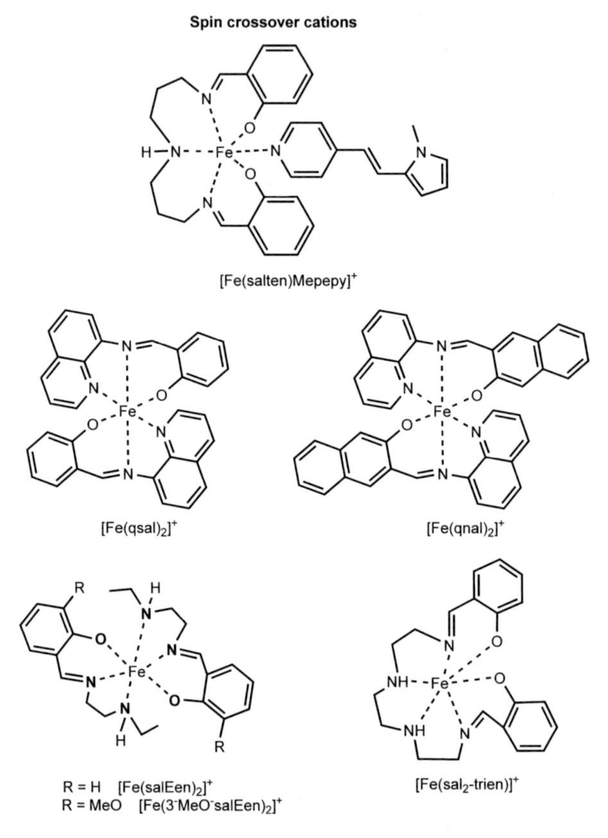

Fig. 8 Spin crossover cations cited in text

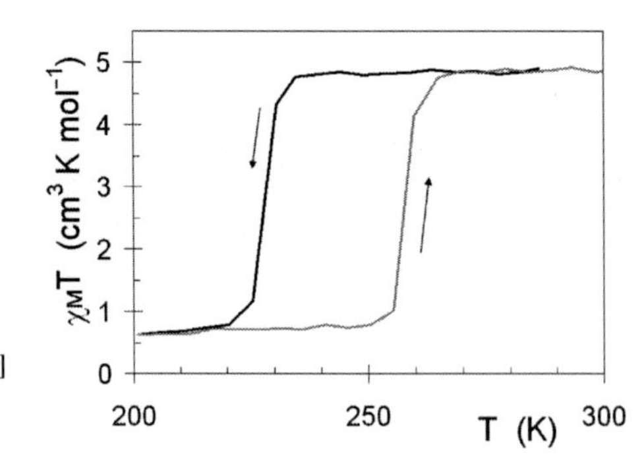

Fig. 9 Magnetic susceptibility measurements for $[Fe(sal_2\text{-}trien)][Ni(dmit)_2]$ showing the thermal hysteresis loop

In the same line, $[Fe(qsal)_2][Ni(dmit)_2].2CH_3CN$ has also been proven to show a cooperative spin transition based on π–π interactions between ligands [105]. Moreover, this compound is also one of the rare Fe^{III} complex to exhibit the light-induced excited spin state trapping (LIESST) effect.

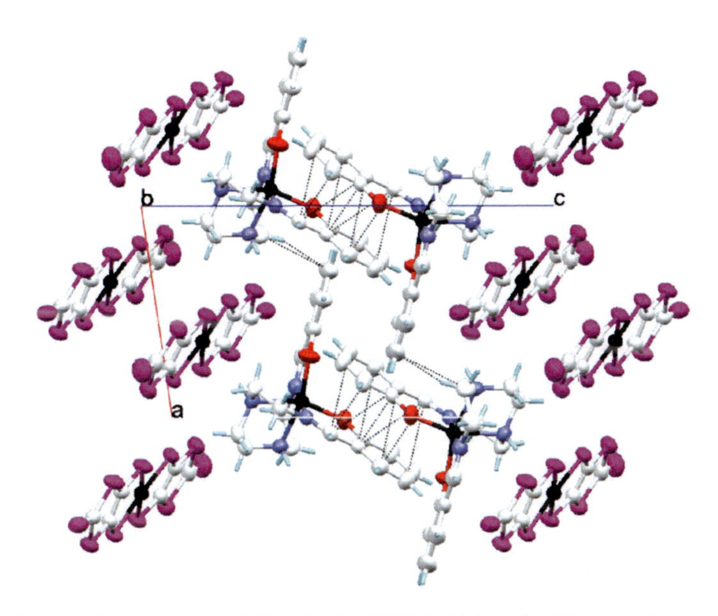

Fig. 10 Structural arrangement of $[Fe(sal_2\text{-trien})][Ni(dmit)_2]$ in the HS state, showing the π–π interactions (*dotted lines*) between $[Fe(sal_2\text{-trien})]^+$ cations

At this stage, the important (but still not well-understood) role of solvent in SCO compounds properties should be pointed out: indeed, the nonsolvated $[Fe(qsal)_2]$ $[Ni(dmit)_2]$ compound was reported 2 years later [103]. The latter does not exhibit any cooperative spin-transition, probably due to the lack of π-interactions observed in its structural arrangement. Therefore, it seems that, whatever the counter-ion, the "control" of π-stacking is one of the key points to enhance cooperativity in SC complexes.

Although combined with $[Ni(dmit)_2]^-$, none of the above-mentioned compounds exhibit electrical properties, since they are 1:1 materials without any charge transfer. One of the first attempts to obtain fractional oxidation state compound was performed by us in 2006 [101]. $[Fe(sal_2\text{-trien})][Ni(dmit)_2]_3$ has been obtained by electrocrystallization from an acetonitrile solution of $[Fe(sal_2\text{-trien})][Ni(dmit)_2]$. This compound behaves like an SC ($\sigma_{RT} = 0.1$ S cm^{-1}) but does not exhibit any spin transition. This seems to be due to the statistical disorder of the whole Fe^{III} complexes, which prevents the occurrence of short contacts between cations.

Chronologically, Sato and coworkers have been the first to obtain and to characterized unambiguously a fractional oxidation state compound containing the Ni $(dmit)_2$ unit and an SCO cation, namely $[Fe(qsal)_2][Ni(dmit)_2]_3.CH_3CN.H_2O$ [106]. This complex has been obtained after electrocrystallization from an acetonitrile solution of $[Fe(qsal)_2][Ni(dmit)_2].2CH_3CN$.

In this compound, $Ni(dmit)_2$ molecules are arranged in columns and $Fe(qsal)_2$ cations are dimerized by π–π interactions and construct a one-dimensional chain. $Fe(qsal)_2$ chains are then interwoven with the $Ni(dmit)_2$ columns (Fig. 11).

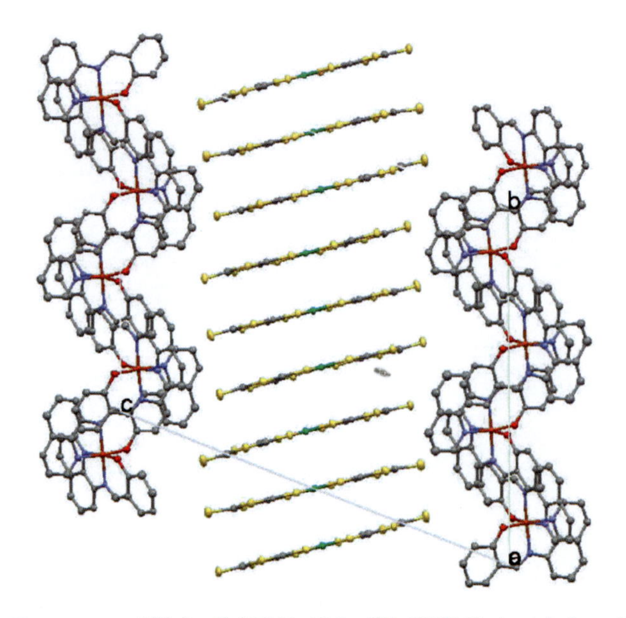

Fig. 11 Crystal structures of $[Fe(qsal)_2][Ni(dmit)_2]_3$. $CH_3CN.H_2O$ viewed along the *a* axis. The solvent molecules were omitted for clarity

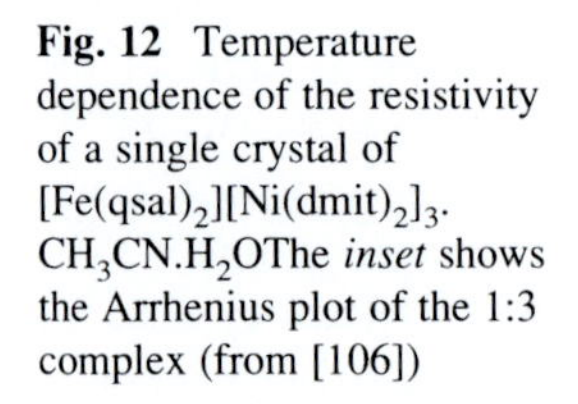

Fig. 12 Temperature dependence of the resistivity of a single crystal of $[Fe(qsal)_2][Ni(dmit)_2]_3$. $CH_3CN.H_2O$The *inset* shows the Arrhenius plot of the 1:3 complex (from [106])

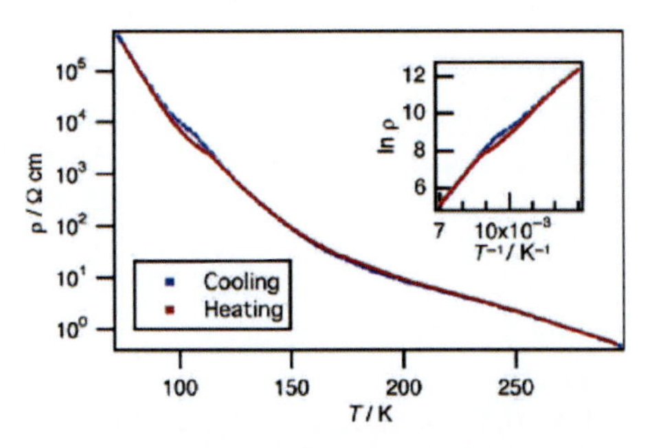

$[Fe(qsal)_2][Ni(dmit)_2]_3.CH_3CN.H_2O$ exhibits a room temperature conductivity of 2.0 S cm^{-1} and behaves as a semiconductor between 300 and 70 K (Fig. 12). It also exhibits a broad spin transition between 300 and 60 K (Fig. 13). Anomalies are observed in the range 120–90 K, both in the magnetic and electric measurements, clearly indicating a synergy between the SCO phenomenon and the electrical conduction.

This compound is also remarkable since it also exhibits a LIESST effect. As claimed by the authors, $[Fe(qsal)_2][Ni(dmit)_2]_3.CH_3CN.H_2O$ can indeed be classified as "a prototypal photoswitchable SCO molecular conductor."

The use of another SCO Fe^{III} complex $[Fe(salEen)_2]^+$ has afforded the fractional oxidation state compound $[Fe(salEen)_2]_2[Ni(dmit)_2]_5.6CH_3CN$ [104]. The latter is

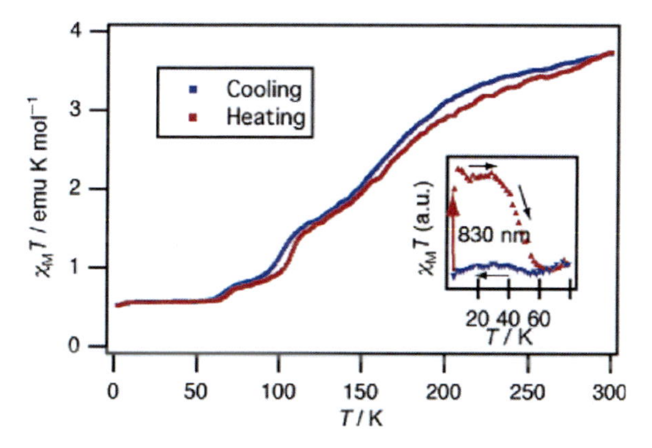

Fig. 13 $\chi_M T$ vs T plot of. [Fe(qsal)$_2$][Ni(dmit)$_2$]$_3$.CH$_3$CN.H$_2$O. The *inset* shows the LIESST effect of the 1:3 complex (from [106])

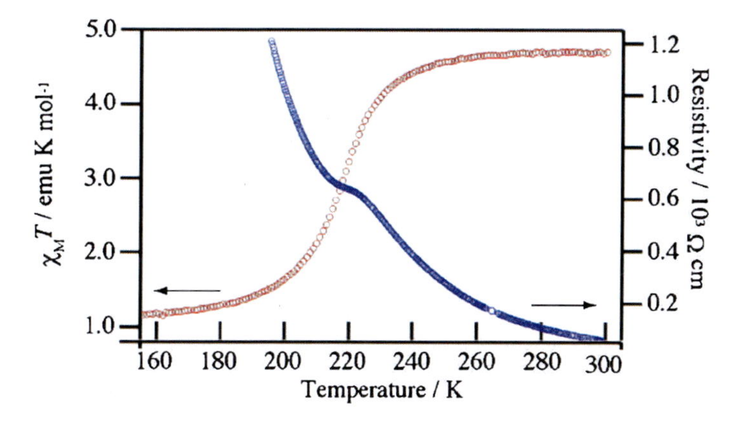

Fig. 14 Temperature dependences of $\chi_M T$ and resistivity for [Fe(qnal)$_2$][Pd(dmit)$_2$]$_5$.(CH$_3$)$_2$CO (from [107])

an SC with a conductivity of 0.12 S cm^{-1} at RT. It also exhibits a complete spin transition between 350 and 50 K. It is then the second SCO molecular conductor.

More recently, [Fe(qnal)$_2$][Pd(dmit)$_2$]$_3$.(CH$_3$)$_2$CO has been reported by Takahashi et al. [107]. The [Fe(qnal)$_2$]$^+$ complex was chosen in order to increase the π–π interactions between units, since the ligand H$_2$qnal contains an additional aromatic 6-ring compared to the ligand H$_2$qsal. This strategy has been proven to be successful since short π–π interactions are observed within the structure of [Fe(qnal)$_2$][Pd (dmit)$_2$]$_3$.(CH$_3$)CO. Moreover, this complex exhibits a spin transition between 240 and 180 K together with an anomaly of the resistivity curve at 220 K, indicating a coupling between the electrical conducting modulation and the spin transition (Fig. 14).

6 Conclusion

Among the large family of metal bis-dithiolenes, the $M(dmit)_2$ family is the only one that has given rise to superconductive materials. It was therefore highly interesting to orientate research tasks to the combination of these complexes with magnetic ones in order to reach bifunctional materials. Whatever the magnetic building blocks reviewed in this chapter, none of the isolated phases exhibits metallic behavior. The conductivity of most phases is poor, mostly because they do not contain fractional oxidation state moieties of the $M(dmit)_2$ building blocks. Moreover, their magnetic properties do not show long range effects. Nevertheless, a few results have shown it is possible to combine SCO and electrical properties, (see $[Fe(qsal)_2][Ni(dmit)_2]_3.CH_3CN.H_2O$ and $Fe(qnal)_2][Pd(dmit)_2]_3.(CH_3)_2CO)$. These results are very encouraging, although the ultimate objective, of efficient reciprocal interactions of conductivity and magnetism in this class of materials remains to be reached.

References

1. Knop W (1842) Justus Liebigs Ann Chem 43:111
2. Zeller HR (1972) Phys Rev Lett 28:1452 LP
3. Zeller HR, Beck A (1974) J Phys Chem Solids 35:77
4. Peierls RE (1952) Quantum theory of solids. Oxford University Press, Oxford
5. Coleman LB, Cohen MJ, Sandman DJ, Yamagishi FG, Garito AF, Heeger AJ (1973) Solid State Commun 12:1125
6. Ferraris J, Cowan DO, Walatka V, Perlstein JH (1973) J Am Chem Soc 95:948
7. Yamada J, Sugimoto T (2004) TTF chemistry. Springer, Berlin Heidelberg New York
8. Saito G, Yoshida Y (2007) Bull Chem Soc Jpn (Commemorative Accounts) 80:1
9. Bechgaard K, Carneiro K, Olsen M, Rasmussen FB, Jacobsen CS (1981) Phys Rev Lett 46:852
10. Kobayashi H, Cui HB, Kobayashi A (2004) Chem Rev 104:5265
11. Robertson N, Cronin L (2002) Coord Chem Rev 227:93
12. Faulmann C, Cassoux P (2004) In: Stiefel EI (ed) Dithiolene chemistry. Synthesis, properties, and applications (Progress in Inorganic Chemistry), vol 52. Wiley, Hoboken, NJ, p 399
13. Kato R (2004) Chem Rev 104:5319
14. Kobayashi A, Fujiwara E, Kobayashi H (2004) Chem Rev 104:5243
15. Alcácer L, Novais H, Pedroso F, Flandrois S, Coulon C, Chasseau D, Gaultier J (1980) Solid State Commun 35:945
16. Steimecke G, Sieler H-J, Kirmse R, Hoyer E (1979) Phosphorus Sulfur 7:49
17. Cassoux P, Valade L (1996) In: Bruce DW, O'Hare D (eds) Inorganic materials. Wiley, Chichester, p 1
18. Cassoux P (1999) Coord Chem Rev 185/186:213
19. Brossard L, Ribault M, Bousseau M, Valade L, Cassoux P (1986) C R Acad Sci Paris 302-II:205
20. Brossard L, Ribault M, Valade L, Cassoux P (1986) Physica B 143:378
21. Tanaka H, Okano Y, Kobayashi H, Suzuki W, Kobayashi A (2001) Science 291:285
22. Murata T, Balodis K, Saito G (2008) Synth Met 158:497
23. Ouahab L, Enoki T (2004) Eur J Inorg Chem 933

24. Hervé K, Gal YL, Ouahab L, Golhen S, Cador O (2005) Synth Met 153:461
25. Kubo K, Nakao A, Ishii Y, Yamamoto T, Tamura M, Kato R, Yakushi K, Matsubayashi G-e (2008) Inorg Chem 47:5495
26. Coronado E, Day P (2004) Chem Rev 104:5419
27. Enoki T, Miyazaki A (2004) Chem Rev 104:5449
28. de Caro D, Faulmann C, Valade L (2007) Chem Eur J 13:1650
29. Gütlich P, Goodwin HA (2004) Spin crossover in transition metal compounds I, II and III. Springer, Berlin Heidelberg New York
30. Brossard L, Ribault M, Valade L, Cassoux P (1989) J Phys 50:1521
31. Brossard L, Hurdequint H, Ribault M, Valade L, Legros JP, Cassoux P (1988) Synth Met 27:157
32. Tajima H, Inokuchi M, Kobayashi A, Ohta T, Kato R, Kobayashi H, Kuroda H (1993) Chem Lett 1235
33. Kobayashi H, Bun K, Naito T, Kato R, Kobayashi A (1992) Chem Lett 21:1909
34. Kato R, Kashimura Y, Aonuma S, Hanasaki N, Tajima H (1998) Solid State Commun 105:561
35. Kato R, Tajima A, Nakao A, Tamura M (2006) J Am Chem Soc 128:10016
36. Kobayashi A, Kim H, Sasaki Y, Kato R, Kobayashi H, Moriyama S, Nishio Y, Kajita K, Sasaki W (1987) Chem Lett 1819
37. Kobayashi A, Kim H, Sasaki Y, Moriyama S, Nishio Y, Kajita K, Sasaki W, Kato R, Kobayashi H (1988) Synth Met 27B:339
38. Kobayashi A, Kobayashi H, Miyamoto A, Kato R, Clark RA, Underhill AE (1991) Chem Lett 20:2163
39. Kato R, Tajima N, Tamura M, Yamaura J-I (2002) Phys Rev B 66:020508
40. Brossard L, Ribault M, Valade L, Cassoux P (1990) Phys Rev B 42:3935
41. Canadell E, Ravy S, Pouget JP, Brossard L (1990) Solid State Commun 75:633
42. Tajima H, Inokuchi M, Ikeda S, Arifuku M, Ohta T, Tamura M, Kobayashi A, Kato R, Naito T, Kobayashi H, Kuroda H (1995) Synth Met 70:1035
43. Kobayashi A, Sato A, Kawano K, Naito T, Kobayashi H, Watanabe T (1995) J Mater Chem 5:1671
44. McConnell HM (1967) Proc Robert A. Welch Found Conf Chem Res 11:144
45. Miller JS, Calabrese JC, Epstein AJ (1989) Inorg Chem 28:4230
46. Broderick WE, Thompson JA, Day EP, Hoffman BM (1990) Science 249:401
47. Broderick WE, Thompson JA, Hoffman BM (1991) Inorg Chem 30:2958
48. Broderick WE, Thompson JA, Godfrey MR, Sabat M, Hoffman BM, Day EP (1989) J Am Chem Soc 111:7656
49. Faulmann C, Pullen AE, Riviere E, Journaux Y, Retailleau L, Cassoux P (1999) Synth Met 103:2296
50. Faulmann C, Riviere E, Dorbes S, Senocq F, Coronado E, Cassoux P (2003) Eur J Inorg Chem 2880
51. Mukai K, Yanagimoyo M, Tanaka S, Mito M, Kawae T, Takeda K (2003) J Phys Soc Jpn 72:2312
52. Mukai K, Matsubara M, Hisatou H, Hosokoshi Y, Inoue K, Azuma N (2002) J Phys Chem B 106:8632
53. Mukai K, Shimobe Y, Jamali JB, Achiwa N (1999) J Phys Chem B 103:10876
54. Mito M, Nakano H, Kawae T, Hitaka M, Takagi S, Deguchi H, Suzuki K, Mukai K, Takeda K (1997) J Phys Soc Jpn 66:2147
55. Mukai K, Konishi K, Nedachi K, Takeda K (1996) J Phys Chem 100:9658
56. Tomiyoshi S, Yano T, Azuma N, Shoga M, Yamada K, Yamauchi J (1994) Phys Rev B Condens Matter Mater Phys 49:16031
57. Kremer RK, Kanellakopulos B, Bele P, Brunner H, Neugebauer FA (1994) Chem Phys Lett 230:255
58. Mukai K, Wada N, Jamali JB, Achiwa N, Narumi Y, Kindo K, Kobayashi T, Amaya K (1996) Chem Phys Lett 257:538

59. Mito M, Takeda K, Mukai K, Azuma N, Gleiter MR, Krieger C, Neugebauer FA (1997) J Phys Chem B 101:9517
60. Takeda K, Hamano T, Kawae T, Hidaka M, Takahashi M, Kawasaki S, Mukai K (1995) J Phys Soc Jpn 64:2343
61. Oakley RT (1988) Prog Inorg Chem 36:299
62. Awaga K, Tanaka T, Shirai T, Fujimori M, Suzuki Y, Yoshikawa H, Fujita W (2006) Bull Chem Soc Jpn 79:25
63. Rawson JM, Banister AJ, Lavender I (1995) Adv Heterocyc Chem 62:137
64. Rawson JM, Alberola A, Whalley A (2006) J Mater Chem 16:2560
65. Rawson JM, Palacio F (2001) Struct Bond 100:93
66. Rawson JM, McManus GD (1999) Coord Chem Rev 189:135
67. Rawson JM, Alberola A, Whalley A (2006) J Mater Chem 16:2560
68. Preuss KE (2007) Dalton Trans 2357–2369
69. Nakatsuji S, Anzai H (1997) J Mater Chem 7:2161
70. Sugimoto H, Aota H, Harada A, Morishima Y, Kamachi M, Mori W, Kishita M, Ohmae N, Nakano M, Sorai M (1991) Chem Lett 2095
71. Kobayashi TC, Takiguchi M, Hong C, Okada A, Amaya K, Kajiwara A, Harada A, Kamachi M (1996) J Phys Soc Jpn 65:1427
72. Ohmae N, Kajiwara A, Miyazaki Y, Kamachi M, Sorai M (1995) Thermochim Acta 267:435
73. Kajiwara A, Mori W, Sorai M, Yamaguchi K, Kamachi M (1995) Mol Cryst Liq Cryst Sci Technol A Mol Cryst Liq Cryst 272:289
74. Kobayashi TC, Takiguchi M, Hong CU, Amaya K, Kajiwara A, Harada A, Kamachi M (1995) J Magn Magn Mater 140/144:1447
75. Kobayashi T, Takiguchi M, Amaya K, Sugimoto H, Kajiwara A, Harada A, Kamachi M (1993) J Phys Soc Jpn 62:3239
76. Kamachi M, Sugimoto H, Kajiwara A, Harada A, Morishima Y, Mori W, Ohmae N, Nakano M, Sorai M et alet al. (1993) Mol Cryst Liq Cryst Sci Technol A Mol Cryst Liq Cryst 232:53
77. Togashi K, Imachi R, Tomioka K, Tsuboi H, Ishida T, Nogami T, Takeda N, Ishikawa M (1996) Bull Chem Soc Jpn 69:2821
78. Lemaire H, Rey P, Rassat A, de Combarieu A, Michel J-C (1968) Mol Phys 14:201
79. Imai H, Inabe T, Otsuka T, Okuno T, Awaga K (1996) Phys Rev B Condens Matter 54: R6838
80. Imai H, Otsuka T, Naito T, Awaga K, Inabe T (1999) J Am Chem Soc 121:8098
81. Aonuma S, Casellas H, Faulmann C, Garreau de Bonneval B, Malfant I, Cassoux P, Lacroix PG, Hosokoshi Y, Inoue K (2001) J Mater Chem 11:337
82. Aonuma S, Casellas H, Bonneval BGd, Faulmann C, Malfant I, Cassoux P, Hosokoshi Y, Inoue K (2002) Mol Cryst Liq Cryst 380:263
83. Aonuma S, Casellas H, Faulmann C, Bonneval BGd, Malfant I, Lacroix PG, Cassoux P, Hosokoshi Y, Inoue K (2001) Synth Met 120:993
84. Mukai K, Hatanaka T, Senba N, Nakayashiki T, Misaki Y, Tanaka K, Ueda K, Sugimoto T, Azuma N (2002) Inorg Chem 41:5066
85. Mukai K, Senba N, Hatanaka T, Minakuchi H, Ohara K, Taniguchi M, Misaki Y, Hosokoshi Y, Inoue K, Azuma N (2004) Inorg Chem 43:566
86. Mukai K, Shiba D, Mukai K, Yoshida K, Hisatou H, Ohara K, Hosokoshi Y, Azuma N (2005) Polyhedron 24:2513
87. Rahman B, Kanbara K-i, Akutsu H, Yamada J-i, Nakatsuji Si (2007) Polyhedron 26:2287
88. Dias MC, Stumpf HO, Sansiviero MTC, Pernaut JM, Matencio T, Knobel M, Cangussu D (2007) Quim Nova 30:904
89. Staniland SS, Fujita W, Umezono Y, Awaga K, Clark SJ, Cui HB, Kobayashi H, Robertson N (2005) Chem Commun 3204
90. Dagotto E, Rice TM (1996) Science 271:618
91. Scalapino DJ (1995) Nature 337:12
92. Hiroi Z, Takano M (1995) Nature 337:41

93. Real JA, Gaspar AB, Munoz MC (2005) Dalton Trans 2062
94. Real JA, Gaspar AB, Niel V, Munoz MC (2003) Coord Chem Rev 236:121
95. Kahn O, Martinez-Jay C (1998) Science 279:44
96. Muller RN, Vander Elst L, Laurent S (2003) J Am Chem Soc 125:8405
97. Bousseksou A, Molnar G, Demont P, Menegotto J (2003) J Mater Chem 13:2069
98. Kunkeler PJ, van Koningsbruggen PJ, Cornelissen JP, van der Horst AN, van der Kraan AM, Spek AL, Haasnoot JG, Reedijk J (1996) J Am Chem Soc 118:2190
99. Dorbes S, Valade L, Real JA, Faulmann C (2005) Chem Commun 69
100. van Koningsbruggen PJ, Maeda Y, Oshio H (2004) Top Curr Chem 233:259
101. Faulmann C, Dorbes S, Real JA, Valade L (2006) J Low Temp Phys 142:265
102. Faulmann C, Dorbes S, Garreau de Bonneval B, Molnar G, Bousseksou A, Gomez-Garcia CJ, Coronado E, Valade L (2005) Eur J Inorg Chem 3261
103. Faulmann C, Dorbes S, Lampert S, Jacob K, Garreau de Bonneval B, Molnar G, Bousseksou A, Real JA, Valade L (2007) Inorg Chim Acta 360:3870
104. Faulmann C, Jacob K, Dorbes S, Lampert S, Malfant I, Doublet M-L, Valade L, Real JA (2007) Inorg Chem 46:8548
105. Takahashi K, Cui H, Kobayashi H, Yasuaki E, Sato O (2005) Chem Lett 34:1240
106. Takahashi K, Cui H-B, Okano Y, Kobayashi H, Einaga Y, Sato O (2006) Inorg Chem 45:5739
107. Takahashi K, Cui H-B, Okano Y, Kobayashi H, Mori H, Tajima H, Einaga Y, Sato O (2008) J Am Chem Soc 130:6688

Top Organomet Chem (2009) 27: 161–189

Magnetic Properties of Radical, Crystalline Mixed Cyclopentadienyl/Dithiolene Complexes

Marc Fourmigué

Abstract This chapter describes in detail a class of heteroleptic complexes associating cyclopentadienyl (Cp) and dithiolene (dt) ligands, essentially known with the $Cp_2M(dt)$, $CpM(dt)_2$, $[CpM(dt)]_2$ and CpM(dt) stoichiometries. The three first classes are reported with early transition metals (groups 4–7), while the latter is restricted to group 9 and 10 metal centers. Depending on the metal and the complex oxidation state, paramagnetic species can be isolated, as for example the formally d^1 species $Cp_2V(dt)$, $[Cp_2Mo(dt)]^+$ or $[CpMo(dt)_2]$, or the formally d^7 CpNi(dt). They all exhibit a rich structural chemistry and variable spin density distribution between Cp, M and dt fragments. Their magnetic properties in the solid state are a combination of an original spin density distribution and specific intermolecular interactions in the crystalline phase. They lead to a variety of behaviors, from independent spins with their associated Curie-type law, singlet-triplet systems and their extension to alternated spin chains and spin ladders, uniform spin chains or ordered antiferromagnetic ground states. Besides the S•••S and S•••M interactions which control most of the antiferromagnetic interactions observed so far in these magnetic organometallic complexes, the contribution of direct Cp•••Cp and Cp•••S interactions is shown to control in some cases the magnetic behavior of the complexes, an original feature in organometallic chemistry.

Keywords Dithiolene complexes, Intermolecular interactions, Metallocenes, Molecular magnetism

Contents

M. Fourmigué
Sciences Chimiques de Rennes, Université Rennes 1 and CNRS UMR 6226, Campus de Beaulieu, 35042, Rennes, France, E-mail: marc.fourmigue@univ-rennes1.fr

M. Fourmigué and L. Ouahab (eds.), *Conducting and Magnetic Organometallic Molecular Materials*, Topics in Organometallic Chemistry 27,
DOI: 10.1007/978-3-642-00408-7_7,

Abbreviations

SOMO Singly occupied molecular orbital
TCNQ Tetracyanoquinodimethane
TMTSF Tetramethyltetraselenafulvalene
TTF Tetrathiafulvalene

1 Introduction

The history of conducting organic materials is punctuated by some landmarks such as the discovery of conducting TTF salts [1] and the metallic character in TTF•TCNQ in the 1970s [2], the superconductivity in the Bechgaard salts derived from tetramethyltetraselenafulvalene (TMTSF) in the 1980s [3], its extension to two-dimensional systems soon after [4] and to three-dimensional systems such as K_3C_{60} by the end of the 1990s [5]. These organic metals are based on radical cation or radical anion salts of electroactive molecules, which strongly interact in the crystalline state to allow for the formation of partially filled energy bands, hence their possible metallic conductivity [6]. Parallel to the work on tetrathiafulvalene derivatives, radical dithiolene complexes were also investigated during the same period for their possible conducting or magnetic properties in the solid state [7, 8].

Dithiolene complexes are based on conjugated 1,2-dithiolate ligands (see Scheme 1 for the different acronyms used to describe the most common dithiolene ligands). The redox active character of this so-called "noninnocent" ligand confers to its metal complexes the possibility of existing in various oxidation states [9], some of them paramagnetic and therefore susceptible to afford electric and magnetic properties in the solid state, by analogy with a $TTF^{+\bullet}$ radical cation or a

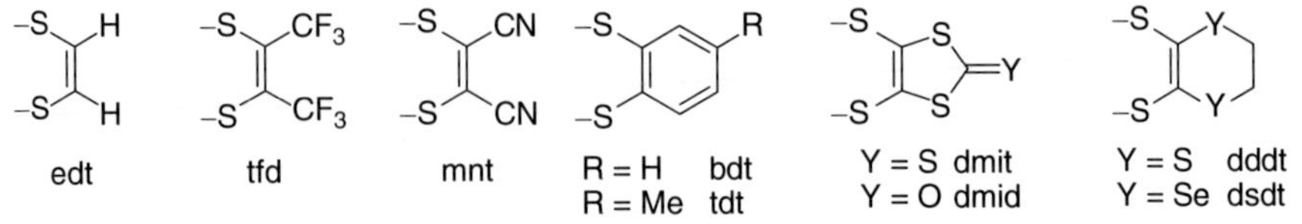

edt | tfd | mnt | R = H bdt, R = Me tdt | Y = S dmit, Y = O dmid | Y = S dddt, Y = Se dsdt

Scheme 1 Acronyms of the most usual dithiolene ligands

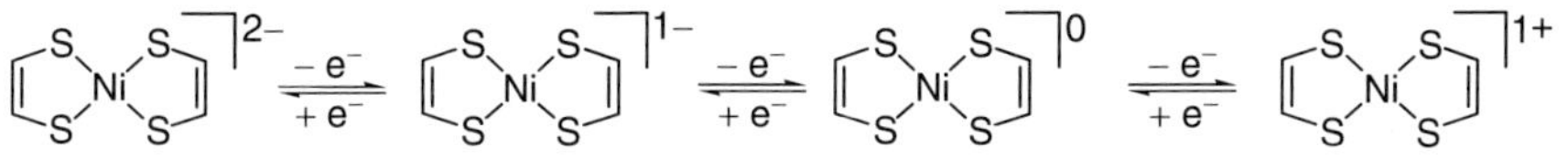

Scheme 2 Different redox states of square–planar nickel bis(edt) complexes

$TCNQ^{-\bullet}$ radical anion. For example, depending on the nature of the substituents R on the dithiolate ligand (Scheme 2), the $[Ni(edt)_2]$ complex can be isolated as dianionic $[Ni(edt)_2]^{2-}$, monoanionic $[Ni(edt)_2]^{-}$, neutral $[Ni(edt)_2]^{0}$ and monocationic $[Ni(edt)_2]^{+}$ complexes.

Among them, both the monoanion $[Ni(S_2C_2R_2)_2]^{-}$ and the monocation $[Ni(S_2C_2R_2)_2]^{+}$ are paramagnetic with $S = 1/2$. They correspond to a formal d^7 Ni(III) and d^5 Ni(V) oxidation state, but their exact electronic description is based on a strong contribution of the dithiolene ligands in the SOMOs [10, 11]. Because of this strong spin density delocalization, intermolecular interactions in the solid state can develop into one-, two- and three-dimensional electronic structures, with their associated magnetic behavior. Furthermore, in the presence of mixed-valence systems, partially filled conduction bands and metallic or even superconducting behavior can be observed. Several excellent review papers are available which described in details these materials based on dithiolene complexes [7, 8, 12]. See also in this book the chapters "New Molecular Architecture for Electrically Conducting Materials Based on Unsymmetrical Organometallic-Dithiolene Complexes" and "Metallocenium Salts of Transition Metal Bis-Dichalcogenate Anions; Structure and Magnetic Properties" by Kato and by Almeida respectively. However, they are essentially limited to the homoleptic square-planar complexes of Cu, Au and the Ni, Pd, Pt triad. This is all the more surprising since, besides the homoleptic bis(dithiolene) complexes described above, tris- [13, 14] and even tetrakis(dithiolene) [15] complexes are also known, which also exhibit rich structural and electrochemical properties. In most cases, however, their magnetic properties in the solid state have not been investigated in detail, probably also because of the limited extent of possible intermolecular interactions [16].

Furthermore, besides these homoleptic complexes, 1, 2-dithiolate ligands can share a metal coordination sphere with a variety of other ligands, phosphines and diphosphines, α-diimines such as 2, 2′-bipyridines and phenantrolines, cyclopentadienyl, . . . These mixed complexes have been extensively investigated, for example for their luminescence properties [17] or as models for enzymes which incorporate mono- or bis(dithiolene) [18, 19]. Complexes which associate both dithiolate and cyclopentadienyl ligands were described for the first time in 1963 by King from the reaction of the bis(trifluoromethyl)dithietene $S_2C_2(CF_3)_2$ with cyclopentadienyl-metal carbonyl to afford $[CpM(tfd)]_n$ (M = V, Cr or Mo, $n = 2$; M = Co or Ni, $n = 1$) [20]. Since then, numerous examples have been reported, with essentially four different Cp/dithiolene ratios (Scheme 3): Cp_2M(dithiolene) and CpM(dithiolene)$_2$ with the early metal atoms (groups 4–7, lanthanides and actinides), dimeric [CpM (dithiolene)]$_2$ composition with group 5 and 6 metals, and monomeric CpM(dithiolene) complexes with group 9–10 metal atoms. A comprehensive review was published on these compounds in 1998 [21].

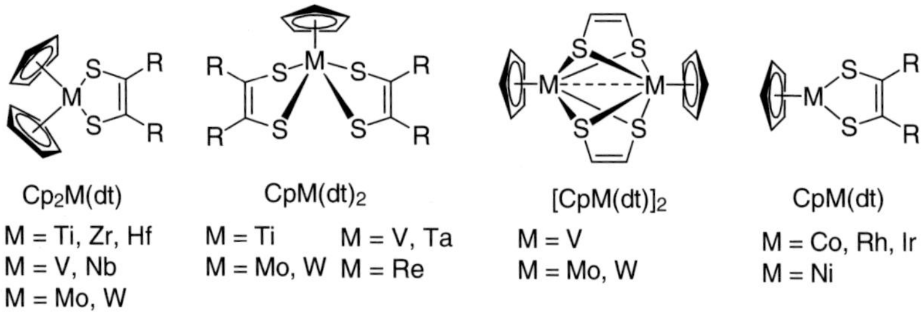

$Cp_2M(dt)$
M = Ti, Zr, Hf
M = V, Nb
M = Mo, W

$CpM(dt)_2$
M = Ti M = V, Ta
M = Mo, W M = Re

$[CpM(dt)]_2$
M = V
M = Mo, W

CpM(dt)
M = Co, Rh, Ir
M = Ni

Scheme 3 The main classes of heteroleptic Cp/dithiolene metal complexes

Among them, however, a very limited number was available in a paramagnetic state and even fewer examples of the nature of their magnetic ground state were described at that time. Over the last 15 years we have described different series of such *paramagnetic* heteroleptic mixed cyclopentadienyl/dithiolene complexes and investigated in detail their magnetic properties in the solid state. Very different behaviors were observed: (1) noninteracting radical species characterized with a Curie law, (2) bimolecular systems with their associated singlet-triplet behavior, (3) interacting dyads into alternated chain or spin ladder systems, (4) three-dimensional network of intermolecular interactions allowing for the setting of an ordered 3D antiferromagnetic ground state [22]. The aim of this chapter is to provide a comprehensive review of the different magnetic behaviors observed in these radical Cp/dithiolene complexes, with specific attention to the relationship between the molecular and solid state structure they adopt on one hand and their magnetic behavior on the other.

2 Radical Heteroleptic Cp/Dithiolene Complexes

As mentioned above, complexes with different Cp/dithiolene ratio have been described in the literature. They will be described successively below, with specific attention to the paramagnetic species available for the investigation of their solid state magnetic properties.

2.1 The Cp_2M(Dithiolene) Complexes

In the Cp_2M(dithiolene) series, d^0 complexes were investigated essentially with Ti, and to a lesser extent with Zr and Hf, in their IV oxidation state. These complexes can be reversibly reduced to the d^1, Ti^{III} anionic species but they were never isolated in the solid state. Attempts to oxidize these d^0 complexes were also unsuccessful, as electrochemical oxidation leads to their decomposition [23, 24]. The essential structural characteristic of these d^0 complexes is the strong folding of

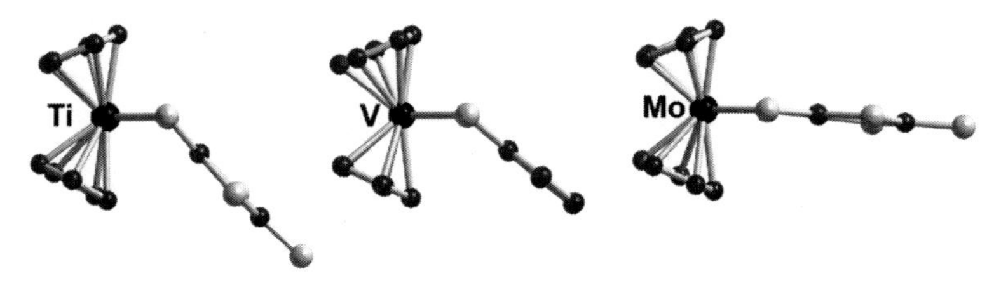

Fig. 1 X-ray crystal structures of Cp_2M(dithiolene) complexes: (**a**) d^0 Cp_2Ti(dmit); (**b**) d^1 Cp_2V (bdt); (**c**) d^2 Cp_2Mo(dmit), showing the evolution of the metallacycle folding angle

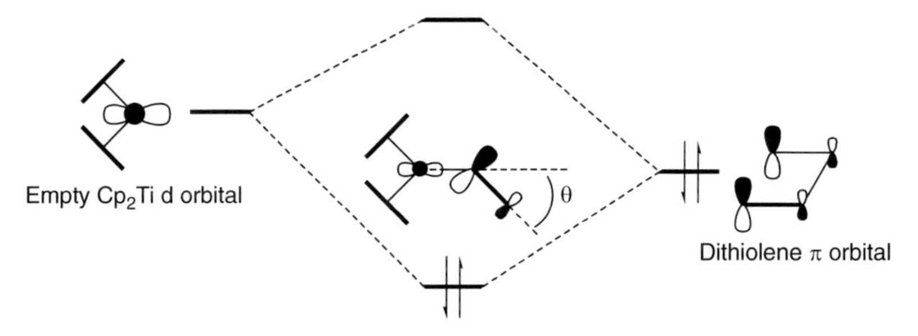

Fig. 2 Simplified electronic interaction diagram between the empty Cp_2Ti and occupied dithiolene fragment orbitals. The interaction is stabilizing only when $\theta \neq 0$

the TiS_2C_2 metallacycle (Fig. 1). It is associated with a favorable mixing of the frontier orbitals of both Cp_2Ti and dithiolene fragments [25, 26]. In the unfolded C_{2v} geometry, the LUMO of the Cp_2Ti fragment is orthogonal to the HOMO, π-type like orbital of the dithiolene fragment. The system can stabilize by folding (Fig. 2), allowing for an orbital mixing with associated structural distortion. The folding angle usually observed amounts to 43–50°.

The radical d^1 Cp_2M(dithiolene) complexes are known with V or Nb in their IV oxidation state. They also exhibit a folding of the MS_2C_2 metallacycle but to a lesser extent than the group 4 metal complexes (34–38°). Indeed, in this situation, the antibonding combination of the two fragment orbitals of the Cp_2M and dithiolene moieties is now singly occupied, while it was empty in the d^0 titanium series; hence the more limited stabilization and folding angle. A folding angle of 38.5° was recently reported in the simplest compound of the series, i.e., Cp_2V(edt) [27]. Note also that the d^0 oxidized cationic complexes $[Cp_2V(dt)]^+$ or $[Cp_2Nb(dt)]^+$ exhibit structural distortions similar to those of the titanium complexes, as evidenced from their variable-temperature NMR behavior [28, 29]. The paramagnetic nature of all d^1 complexes is confirmed by solution EPR spectroscopy. Hyperfine coupling constants, either with ^{51}V (I = 7/2) or with ^{93}Nb (I = 9/2), are comparable to those reported for the analogous 1, 1-dithiolate complexes, demonstrating that the spin density is here essentially localized on the metal. Recently, dithiolene-appended porphyrazines incorporating two S = 1/2 vanadocene moieties were reported with the EPR data pointing for some magnetic interaction through the delocalized porphyrazine core [30, 31].

The neutral diamagnetic d^2 Cp_2M(dithiolene) complexes are known with Mo and W and a variety of dithiolene ligands. As one could anticipate from this electron count, the metallacycle in these 18-electron complexes is not folded any more. They all exhibit reversible oxidations to the formally d^1 and d^0 species. The d^1 $[Cp_2M(dithiolene)]^{+\bullet}$, M = Mo, W, has been investigated in detail. In these radical complexes, the folding of the metallacycle cycle already observed in the d^1 V or Nb neutral complexes at 34–38°, was found to be not only smaller (0–30°) but also highly dependent of the counter ion. For example, the very same complex $[Cp_2Mo(dmit)]^{+\bullet}$ was observed in an unfolded conformation in its PF_6^- salt ($\theta = 0°$) [32], while the MoS_2C_2 metallacycle was folded by 30° in its bromide salt [33]. Theoretical calculations have shown that the energy cost for the folding of these d^1 complexes was less than 1 kcal mol^{-1} for θ between 0 and 35° [27, 34].

We are therefore faced here with radical complexes which easily distort depending on the structural arrangement and whose SOMO is different for every crystal structure associated with a given counter-ion, a very original feature in these series. The unfolded d^1 complexes can be described as Mo(IV) complexes with a spin density essentially localized on the dithiolene ligand while the more folded complexes have a stronger metal character. This variable spin density delocalization is expected to influence strongly the amplitude and dimensionality of intermolecular interactions between radical species in the solid state, as detailed below in Sect. 3.

The analogous uranium(IV) and uranium(III) complexes, Cp^*_2U(dddt) [35] and $[Cp^*_2U(dddt)]^-$ [36] were also recently described and structurally characterized. They both exhibit a strong folding of the metallacycle, 56.3(3) and 51.9(2)°, for the U(IV) and U(III) complexes respectively. The analogous Ce(III) anionic complex $[Cp^*_2Ce(dddt)]^-$ was also prepared and found to be closely related to the U(III) analog. A contraction of the metal-sulfur bond lengths on going from the cerium to the uranium anionic complex is partly related to the uranium $5f$ orbital-ligand mixing which is greater than the cerium $4f$ orbital-ligand mixing. Similarly, a rather low uranium net charge (+0.69) compared to that of cerium (+1.33) in the anionic complexes denotes an increased ligand-to-metal donation. The metal spin densities were found roughly equal to 1 and 3 for the Ce(III) and U(III) complexes, respectively, in good agreement with the free metal ion configurations, which are $4f^1$ for the former and $5f^3$ for the latter. The unrestricted calculations give a metallic spin density equal to 2.29 for the neutral uranium(IV) complex. Such a value obtained for a molecule in its triplet state, for which the number of unpaired electrons is 2, indicates that a small negative spin density is localized on the ligands. However, the magnetic properties of these air-sensitive complexes were not investigated, neither in solution, nor in solid state.

Besides modifications of the metal center in these Cp_2M(dithiolene) complexes, the two cyclopentadienyl rings can also be modified and several examples of paramagnetic d^1 molybdenum and tungsten complexes have been reported (Fig. 3), as rigid *ansa*-metallocene in $[Me_2C(\eta^5\text{-}C_5H_4)_2M(dmit)]^{+\bullet}$, M = Mo, W [37] or with the more sterically demanding *tert*-butyl derivative $[(\eta\text{-}C_5H_4{}^tBu)_2W(dmit)]^{+\bullet}$ [38]. Surprisingly, the *ansa* derivative $[Me_2C(\eta^5\text{-}C_5H_4)_2M(dmit)]^{+\bullet}$ exhibits only a limited folding of the metallacycle, which might be related to a modification of the Cp_2Mo fragment orbitals due to the *ansa* constraint (Fig. 3).

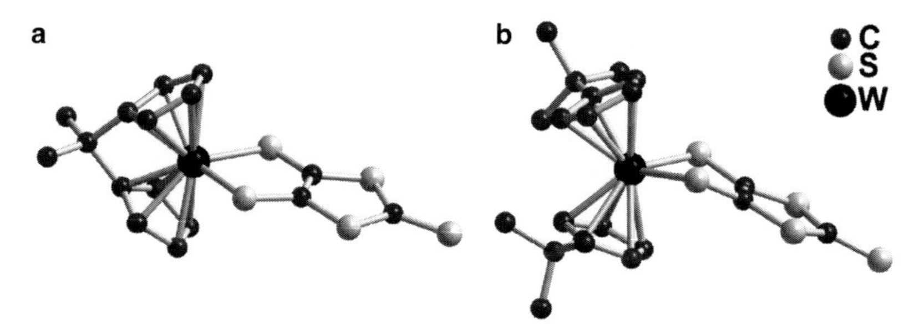

Fig. 3 X-ray crystal structures of the radical tungsten complexes in $[Me_2C(\eta^5\text{-}C_5H_4)_2W(dmit)]^{+\bullet}$ and $[(\eta\text{-}C_5H_4Bu^t)_2W(dmit)]^{+\bullet}$

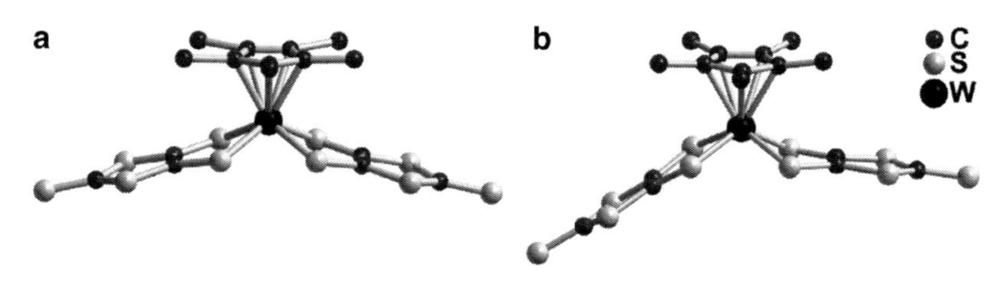

Fig. 4 (**a**) The anionic d^2 $[Cp^*W(dmit)_2]^-$ and (**b**) neutral radical d^1 $[Cp^*W(dmit)_2]^\bullet$ complexes

2.2 *The CpM(Dithiolene)₂ Complexes*

Complexes of general formula CpM(dithiolene)$_2$ were similarly described with group 4–6 metals together with two examples with d^2 Re(V) species, that is Cp*Re(tdt)$_2$ and Cp*Re(dcmedt)$_2$ [39, 40] (dcmedt: 1,2-bis(carboxymethyl)-ethylene-1, 2-dithiolato). The two latter examples were structurally characterized but electrochemical data were not reported. The Ti(IV) anionic complexes $[CpTi(dithiolene)_2]^-$ reduce to the Ti(III) species as their Cp$_2$Ti(dithiolene) analogs. The group 5 complexes such as (MeCp)V(tfd)$_2$ or Cp*Ta(edt)$_2$ are known in their neutral state, that is in a d^0 diamagnetic configuration. On the other hand, the anionic d^2 $[CpW(tfd)_2]^-$ or $[Cp^*M(dmit)_2]^-$ (M = Mo, W) 16-electron complexes exhibit reversible oxidation waves to the d^1 and d^0 species. The radical 15-electron d^1 neutral complexes have indeed been isolated in the solid state, as powder for [CpW(tfd)$_2$] [41], as single crystals for [Cp*Mo(dmit)$_2$] [42] and [Cp*W(dmit)$_2$] [43] (Fig. 4), and more recently for [CpMo(dcmedt)$_2$] [44] and [CpMo(S$_2$C$_2$Ph$_2$)$_2$] [45]. As described above for the Cp$_2$M(dithiolene) complexes (M = Mo, W), variable folding of the metallacycles are observed in the 16- and 15-electron complexes, associated in the latter to a strong delocalization on the dithiolene ligands. It is confirmed by EPR experiments in solution performed on [CpMo(S$_2$C$_2$Ph$_2$)$_2$] where the A_{Mo} coupling amounts to 9.15 G, a small value when compared with classical metal-centered d^1 Mo complexes with A_{Mo} values of 30–50 G [46]. Similarly, the g values obtained from frozen solution for

$[CpMo(S_2C_2Ph_2)_2]$ at 2.0275, 2.0074 and 1.9936 or from oriented single-crystal for $[Cp^*Mo(dmit)_2]$ at 2.027, 2.012 and 1.992, demonstrate an extensive delocalization of the unpaired electron the dithiolene ligands. The solid state properties of these series of radical complexes will be described below in detail in Sect. 3.

Uranium complexes analogous to these compounds were also described, but with the cot (cyclooctatetraene dianion) ligand rather than the Cp or Cp* ones. Both the dianionic U(IV) $[(cot)U(dddt)_2]^{2-}$ [47] and monoanionic U(V) $[(cot)U(dddt)_2]^{-}$ [48] complexes were isolated and structurally characterized (Fig. 5). Spectacular distortions of the US_2C_2 metallacycles were rationalized on the basis of DFT calculations, which reproduced the spectacular folding of the *endo* US_2C_2 metallacycle when the dianionic species undergoes an oxidation.

The calculations also bring to light a significant U•••(C=C) interaction between the metal center and the C=C bond of the *endo* dithiolene ligand in the U(V) anionic complex, which does not exist in the dianionic species. Again, magnetic properties of these air-sensitive compounds were not investigated.

2.3 The Dimeric [CpM(Dithiolene)]₂ Complexes

These dimeric complexes involve, in their neutral state, two metal atoms in the (III) oxidation state. In the vanadium complexes such as $[CpV(bdt)]_2$ and $[CpV(tft)]_2$, the V−V bond length, 2.54 Å in $[CpV(bdt)]_2$, are shorter than observed in model complexes with a single V−V bond, indicating a partial double-bond character, also confirmed by a measured magnetic moment of 0.6 μ_B in $[CpV(tfd)]_2$, lower than expected if the two remaining unpaired electrons contribute to the magnetic susceptibility [20, 49]. This class of complexes most probably deserves deeper attention in order to understand their exact electronic structure.

The Mo(III) d^3–d^3 complexes $[CpMo(dithiolene)]_2$ are characterized by a single Mo−Mo bond, further stabilized by interaction with the π system of the dithiolene ligands. Indeed, the analogous complexes where the two dithiolene are replaced by four thiolate groups were found to oxidize more easily and salts of the cationic d^3–d^2 $[CpMo(SMe)_4MoCp]^+$ were even isolated and structurally and magnetically characterized [50].

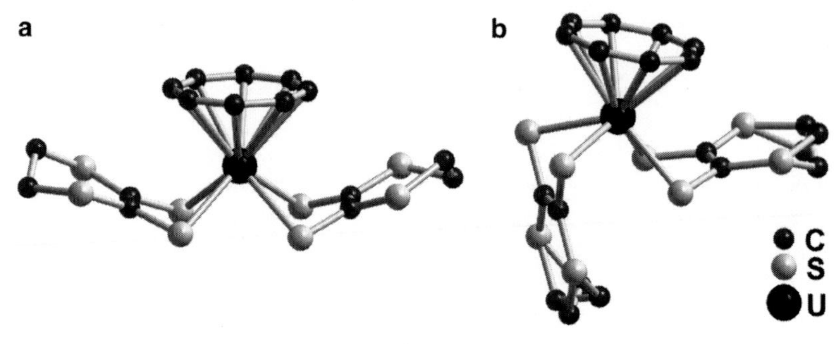

Fig. 5 (**a**) The U(IV) dianionic $[(cot)U(dddt)_2]^{2-}$ (**b**) and the U(V) $[(cot)U(dddt)_2]^{-}$ complexes

2.4 The Monomeric [CpM(Dithiolene)] Complexes

As in the dimeric compounds described above, the [CpM(dithiolene)] complexes in their neutral state are in their (III) oxidation state. Most of them were originally described with the group 9 metals (Co, Rh, Ir) and, accordingly, are diamagnetic [21]. These complexes can be reversibly reduced to the Co(II) anionic species which were characterized by EPR but never isolated in the solid state. One attempt to oxidize such 16-electron complexes to the paramagnetic cation has been reported with [Cp*Co(dddt)] with the hope that its low redox potential due to the combined effect of the electron rich Cp* and dddt ligands would eventually allow the isolation of the oxidized form [51]. However, the cationic 15-electron species dimerizes and a diamagnetic $[Cp^*Co(dddt)]_2^{2+}$ dication was isolated, characterized by very short intradimer Co–S bonds (Fig. 6a). These $[CpCo(dithiolene)]^+$ radical complexes can also be stabilized through Co coordination, either with Lewis bases as trimethylphosphite [51–53], or by halide anions as observed in the bromine oxidation product of [Cp*Co(dmit)] to give [Cp*Co(dmit)Br] [54, 55] (Fig. 6b).

Only two CpNi(dithiolene) complexes were described in our earlier review [21], these being CpNi(tfd) [20, 56] and CpNi(dmit) [57]. Over the last 10 years, we have developed novel synthetic procedures [58], allowing for the isolation and structural characterization of new examples in this class of complexes. Their paramagnetic nature had already been identified by Green in 1963 in [CpNi(tfd)] from the determination of its magnetic moment at room temperature [20]. These formally Ni(III) neutral radical complexes are characterized by an extensive delocalization of the spin density, which also depends on the nature of the dithiolene ligand. For example, the calculated Cp/Ni/dithiolene spin density distribution amounts to 12/28/60% in the CpNi(dddt) complex, compared with 21/37/42% in CpNi(bdt) [59] (Fig. 7). An important point is the high spin density of the Cp ring in the bdt complex, the origin of a strong intermolecular Cp•••Cp interaction in the solid state (see Sect. 3).

An experimental determination of the spin density distribution in these complexes was recently reported, based on EPR data on ^{77}Se-enriched (at 100%) diselenolene analogs where the ^{77}Se coupling constants and g tensors could be determined from frozen solutions spectra [60]. A 14% spin density was determined on each Se atom in ^{77}Se-enriched [CpNi(bds)] while a larger 16% spin density on

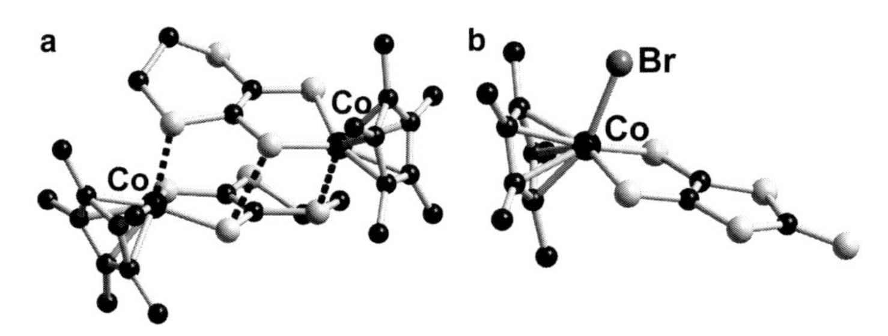

Fig. 6 Stabilization of oxidized $[Cp^*Co(dithiolene)]^+$ complexes through (**a**) dimerization in $[Cp^*Co(dddt)]_2^{2+}$ and (**b**) coordination by bromide anion in Cp*Co(dmit)Br

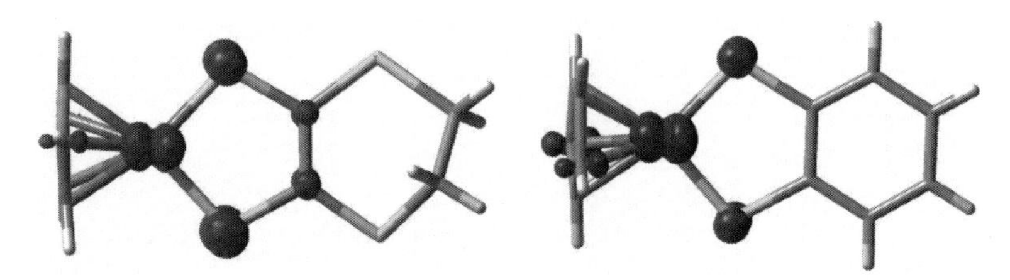

Fig. 7 Spin density distribution in CpNi(dddt) (*left*) and CpNi(bdt) (*right*) (taken from [59], with authorization of the American Chemical Society)

each ^{77}Se atom was found in ^{77}Se-enriched [CpNi(dsit)], in accordance with the DFT calculations.

2.5 Preliminary Conclusions

As exemplified above, among the various heteroleptic $[Cp_nM(dt)_m]$ complexes described so far, only a few series have been isolated in a *radical* state; these are collected in Scheme 4 and finally concern only three classes, according to the formal electron count on the metal center:

1. The (formally) d^1 complexes derived from the early transition metals in $[Cp_2M(dt)]^{n+}$ with $n = 0$, M = V, Nb; $n = 1$, M = Mo, W, or in neutral $[CpM(dt)_2]$ with M = Mo, W
2. The formally d^5 cobalt complexes stabilized by coordination as in cationic $[CpCoL(dt)]^+$ or neutral [CpCoX(dt)]
3. The formally d^7 neutral nickel complexes [CpNi(dt)]

It should also be stressed here that many of these complexes are neutral and therefore relatively soluble in common organic solvents, an important issue for their purification and crystallization. Among all these paramagnetic complexes, only a fraction has been investigated for their magnetic properties in the solid state,

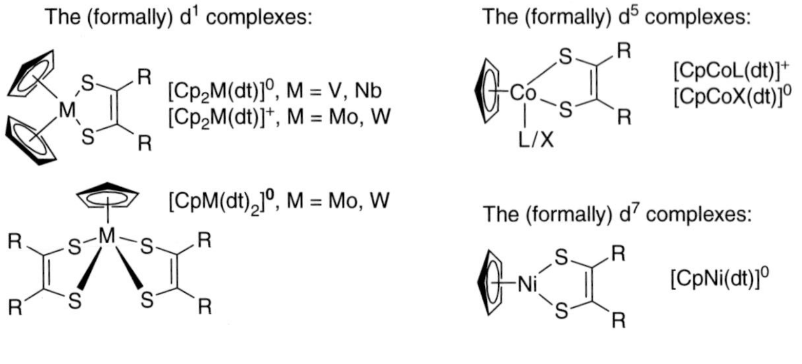

The (formally) d^1 complexes:

$[Cp_2M(dt)]^0$, M = V, Nb
$[Cp_2M(dt)]^+$, M = Mo, W

$[CpM(dt)_2]^0$, M = Mo, W

The (formally) d^5 complexes:

$[CpCoL(dt)]^+$
$[CpCoX(dt)]^0$

The (formally) d^7 complexes:

$[CpNi(dt)]^0$

Scheme 4 Radical $[Cp_nM(dt)_m]$ complexes

essentially through the determination of temperature dependence of their magnetic susceptibility.

If we are concerned with their magnetic properties in the solid state, one of the essential parameters for these organometallic radical complexes involving dithiolene ligands is therefore the degree of delocalization of the spin susceptibility between the metal, the cyclopentadienyl and the dithiolene ligands. In the square-planar homoleptic dithiolene complexes, the radical species are extensively delocalized on the ligands and the intermolecular interactions which control the magnetic and conducting properties are directly related to the number, extend and dimensionality of the S•••S overlap interactions. The paramagnetic [Cp_nM(dithiolene)$_m$] complexes described above differ in many aspects from the homoleptic bis (dithiolene) complexes:

1. They are not planar and cannot stack on top of each other into extended chains with a strong overlap between molecules. As a consequence, there is little chance to stabilize partially filled conduction bands with their associated metallic conductivity.
2. As mentioned above for the [Cp_2Mo(dithiolene)]$^{+\bullet}$, the [Cp*Mo(dithiolene)$_2$]$^{\bullet}$ and the [CpNi(dithiolene)]• radical complexes, the spin density is not only partially delocalized on the dithiolene ligand but also on the metal and even the Cp rings. This peculiar feature opens new paths for intermolecular interactions in the solid state besides the direct dithiolene/dithiolene overlaps, since Cp/dithiolene and Cp/Cp contacts are also to be considered.
3. Most of the conducting or magnetic materials investigated in the homoleptic series are salts while many of the radical heteroleptic complexes described above are in their neutral state, potentially allowing for a stronger three-dimensional character.

3 Magnetic Properties in the Solid State

Most of the complexes described so far in this chapter, because they incorporate such noninnocent ligands as the dithiolene ones, are prone to exhibit intermolecular interactions through dithiolene/dithiolene short contacts in the crystalline solid state. This propensity is further enhanced with those dithiolene ligands such as dmit or dddt because the additional sulfur atoms on the periphery of the metallacycle certainly favors intermolecular interactions through additional S•••S contacts, provided that these outer S atoms bear a non-negligible part of spin density. The evaluation of such intermolecular interactions was essentially based on the evolution of the magnetic susceptibility of the compounds with temperature, combined with an analysis of their solid state structure, and was eventually supported by theoretical calculations of these magnetic interaction energies.

In the following, we have gathered together those complexes which exhibit a similar magnetic behavior, that is: (1) noninteracting radical species characterized with a Curie law, (2) bimolecular systems with their associated singlet-triplet behavior, (3) interacting dyads into alternated chain or spin ladder systems, and (4) three-dimensional network of intermolecular interactions allowing for the setting of an ordered 3D antiferromagnetic ground state. Such an organization is aimed at unraveling and highlighting some specific trends and eventually at giving some indications to anticipate and/or predict possible magnetic state from the molecular geometry of the complexes.

3.1 Isolated Radical Species

The temperature dependence of the molar magnetic susceptibility (χ) of an assembly of paramagnetic spins without interaction is characterized by the Curie behavior with $\chi = C/T$ where $C = Ng^2\beta^2S(S+1)/3k$. It is a very common situation in the organometallic chemistry of radical species when the spin density is essentially localized on the metal atom. Since, in most cases, this atom is surrounded by various "innocent" ligands, intermolecular interactions are very weak and in most cases are reflected by a small contribution described by a Curie-Weiss behavior, with $\chi = C/(T-\theta)$ where θ is the Curie–Weiss temperature. A positive value for θ reflects ferromagnetic interactions while a negative value – the most common situation – reflects an antiferromagnetic interaction.

3.1.1 The Cp_2M(Dithiolene) Series M = V, Nb and Mo

As described above (Sect. 2.1), the EPR data for the d^1 neutral vanadium or niobium complexes indicate a strong spin density on the metal atom with little if any delocalization of the dithiolene ligand. As a consequence, the temperature dependence of the magnetic susceptibility of the neutral radical complexes [Cp_2Nb(dmit)] and [Cp_2Nb(dmid)] follows a Curie-type behavior of noninteracting spins [28, 61]. It is also the case for the ^{t}Bu-Cp derivative, [(η^5-$^tBuC_5H_4)_2$Nb(dmit)] where the Curie behavior was deduced from the spin susceptibility obtained by integration of the solid state EPR signal [28]. Another example of noninteracting Cp_2M(dt) radical complexes is provided by the d^1 [Cp_2Mo(dmit)]$^+$ salt with the bulky [ReO(dmit)$_2$]$^-$ anion [62].

3.1.2 The Uranium Complexes

At variance with the d^1 V or Nb complexes described above, the uranium complexes owe their paramagnetism to 5*f*-electrons. Although these *f* orbitals, and particularly the 5*f* in the actinides when compared with the 4*f* in the lanthanides, interact with the dithiolene orbitals to lead to strong distortions of the US_2C_2 metallacycles (see Fig. 5) [35, 36, 47, 48], the SOMOs themselves are essentially localized on f orbitals without any chance to overlap in the solid state with neighboring molecules. One single example has been reported to date in the $5f^1$ U(V) $[(cot)U(dddt)_2]^-$ isolated as $[Na(18\text{-crown-6}) \bullet THF]^+$salt, where a Curie type behavior was deduced from SQUID magnetic susceptibility measurements [35].

3.2 Dyadic Systems, One-Dimensional Systems

The two-by-two association of radical complexes is the most common motif observed among these series. It follows directly from the general tendency of any radical species to associate into dyads, thus stabilizing the two unpaired electrons into the bonding combination of the two SOMOs. The peculiarities encountered with one or the other complexes are therefore essentially based on: (1) the degree of interactions within the dyad, (2) the nature of the interactions between the dyads in the solid. When the interaction is rather strong, interdyad interactions become negligible and the dyad can be considered as essentially isolated. It is then characterized by a typical magnetic behavior known as singlet-triplet or Bleaney-Bower [63] behavior with the molar susceptibility χ (for one radical species) which reads as

$$\chi = \frac{Ng^2\beta^2}{kT[3 + \exp -J/kT]} \tag{1}$$

where J is referred to as the isotropic interaction parameter and represents the energy difference between the singlet ($S = 0$) and the triplet ($S = 1$) state of the dyad. When the singlet state is the ground state, the interaction is said to be antiferromagnetic and J is then negative. When the triplet state ($S = 1$) is the ground state, the interaction is said to be ferromagnetic and J is then positive. The latter situation has not been described yet in heteroleptic dithiolene complexes and very few examples have been reported among homoleptic dithiolene complexes [64].

Very often, however, these dyads are not isolated in the solid state and interact with neighboring ones at least along one preferential direction. In this case, we can distinguish two important situations: the alternated spin chain and the spin ladder. As shown in Scheme 5, the alternated spin chain is characterized with two different magnetic interactions, noted J and αJ with $0 < \alpha < 1$. Note that if $\alpha = 0$, one recovers the singlet-triplet behavior while if $\alpha = 1$, we are in the presence of a uniform spin chain. The spin ladder is also characterized by two J values, noted $J_{//}$ and $J_{\perp}$ in the following.

There is a fundamental difference between the uniform spin chain on one hand and the alternated spin chain or spin ladder on the other. Indeed, in the latter cases, the ground state is the singlet state and the susceptibility thus goes to zero at the lowest temperatures, with an activated part of the susceptibility between 0 and T (χ_{max}). On the other hand, as shown by Bonner and Fisher [65], in the uniform spin

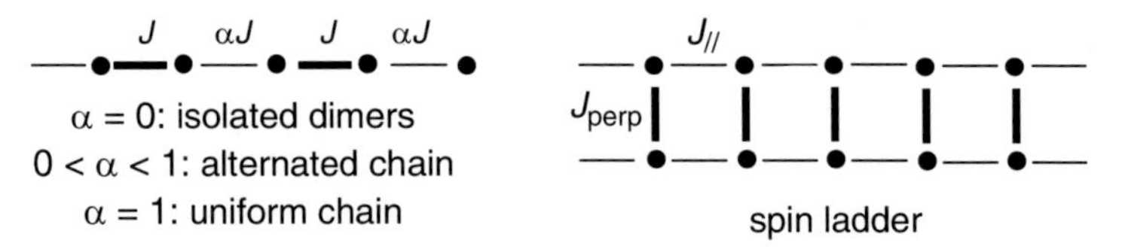

Scheme 5 Models of the spin chain (*left*) and the spin ladder (*right*). The *black dots* represent the spin carriers

chain with antiferromagnetic interaction ($J < 0$), the energy levels form a continuum from one of the $S = 0$ levels up to the unique $S = n/2$ level [66]. At absolute zero, only the bottom of this continuum is thermally populated but, since there is no gap between the $S = 0$ levels and the levels immediately above it, χ does not tend to zero and passes through a rounded maximum at $T(\chi_{max})$ defined by $kT(\chi_{max})/|J| = 0.641$ for a spin Hamiltonian defined as

$$H = -J \sum_{i=1}^{n-1} S_i.S_{i+1}. \tag{2}$$

Analytical expressions of the susceptibility have been developed for the uniform chain [67] as well as for the alternated chain [68].

3.2.1 The Salts of the Cationic d^1 $[Cp_2Mo(dt)]^+$ Complexes

We have already mentioned a very strong dyadic association in the formally d^5 cobalt complexes such as $[Cp^*Co(dddt)]^+$ which dimerizes in the solid state to a fully diamagnetic dicationic dyad (Fig. 6a). It represents the extreme situation where the two radicals form a true $2e^-$ bond, with the sulfur atom of one dithiolene ligand entering the coordination sphere of the other metal. It should be considered as the consequence of the electron deficiency of these cationic $[CpCo(dt)]^+$ 15-electron complexes.

On the other hand, the cationic d^1 $[Cp_2M(dt)]^+$ (M = Mo, W) complexes can be viewed as 17-electron complexes. They do not form such diamagnetic dicationic dyads but often crystallize into centrosymmetric space groups where the cations are associated two by two into inversion-centered dyads. This is typically the case in $[(^tBuCp)_2W(dmit)][TCNQF_4]$ and in $[Cp_2Mo(dmit)][BF_4]$ (Fig. 8). In the former, the bulky *tert*-butyl groups and the $TCNQF_4^-$ anions isolate the dyads from each other, the susceptibility goes through a maximum at $T(\chi_{max})$ of 22 K, leading to $J/k_B = -37.2$ K ($J = -25.8$ cm^{-1}) [38]. In the latter, however [33], a fit of the susceptibility with the Bleaney-Bowers equation together with a Curie-Weiss contribution to take into account a small fraction ρ of magnetic defaults visible at the lowest temperatures was not totally satisfactory and weaker magnetic interactions between dyads were taken into account in a mean-field approximation with

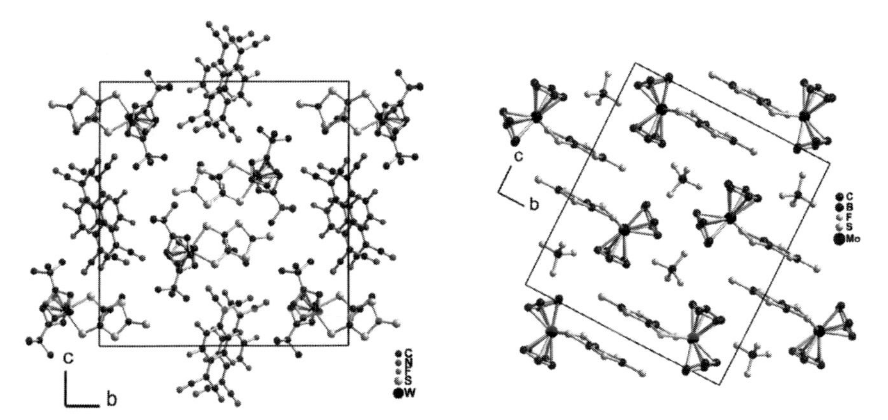

Fig. 8 Projection views of (*left*) $[(^tBuCp)_2W(dmit)][TCNQF_4]$ and (*right*) $[Cp_2Mo(dmit)][BF_4]$ showing the association of the cation into isolated dyads

$$\chi = (1-\rho)\frac{\chi_{dyad}}{1-(4zJ'\chi_{dyad}/Ng^2\mu_B^2)} + \rho\frac{N\mu_B^2}{k_B(T-\theta)} + \chi_{dia}, \tag{3}$$

where ρ is the fraction of magnetic defaults, z is the number of dyads next neighbors (z = 6) and J' is the average of interdyads magnetic exchange constants. An excellent agreement with the experimental data was found that yields $J/k_B = -290$ K (200 cm^{-1}), $J'/k_B = -15$ K, $\rho = 1.4\%$ and $\theta = -2.5$ K.

An extensive series of $TCNQF_4$ salts was also reported with six different complexes, incorporating Mo or W as metal center, dmit, dmid or dsit as dithiolene (or diselenolene) ligand [69]. As already observed above, the dmid salts $[Cp_2M(dmid)][TCNQF_4]$ adopt a recurrent structure with dyads of $(TCNQF_4)_2^{2-}$, alternating with dyads of $[Cp_2M(dmid)]^+$ cations (Fig. 9). With the salts of the dmid complexes $[Cp_2Mo(dmid)][TCNQF_4]$ and $[Cp_2W(dmid)][TCNQF_4]$, the cations form face-to-face dyads which further stack on top of each other leading to the formation of a spin ladder system.

The temperature dependence of the magnetic susceptibility confirms the presence of a gapped system with a singlet ground state. Indeed, in the limit where the exchange coupling $|J_\perp|$ within the rungs is larger than the exchange coupling $|J_{//}|$ along the chains, the ground state has a total spin of $S = 0$ because each rung is in a spin singlet. The system is characterized by a spin gap Δ given by

$$\Delta_{spin} \approx |J_\perp| - |J_{//}| + J_{//}^2/2|J_\perp|. \tag{4}$$

In the low-temperature activated regime, the susceptibility can be written as

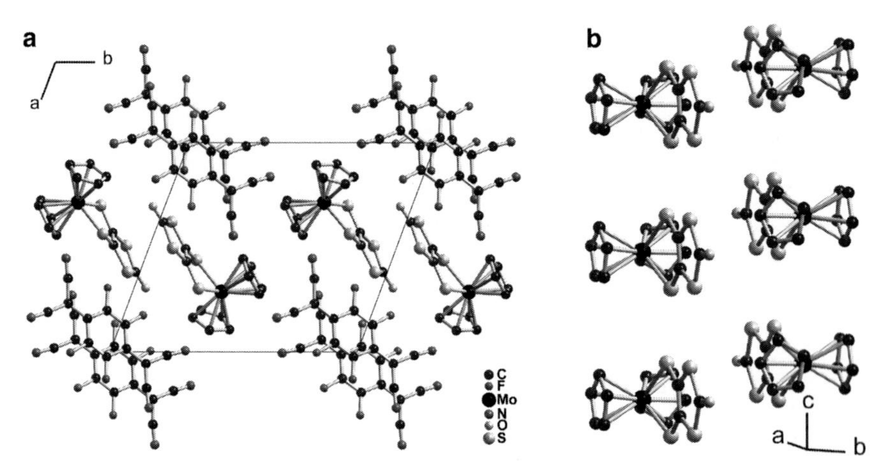

Fig. 9 (**a**) Projection view of the unit cell of $[Cp_2Mo(dmid)][TCNQF_4]$. (**b**) Detail of the ladder-like motif running along *c*

$$\chi \propto T^{-1/2} \exp(-\Delta_{spin}/kT) \tag{5}$$

reflecting the finite spin gap. From the fits, spin gaps Δ_{spin} for the Mo/dmid and W/dmid salts were found at 74 and 13 K respectively; the other magnetic characteristics of the two salts are collected in Table 1, showing that the Mo salt exhibits much stronger intermolecular interactions than the W one, a consequence of a larger part of the spin density on the dmid ligand in the Mo salt. This behavior can be tentatively ascribed to the smaller folding angle of the metallacycle in the Mo complex (21.2°) rather than in the W complex (26.6°). The difference is, however, relatively small and precise determinations of the spin density distribution, from EPR data and/or from theoretical calculations, are needed to ascertain this point.

The dyad geometry was completely different in the $TCNQF_4$ salt of the dsit complexes $Cp_2Mo(dsit)$ and $Cp_2W(dsit)$ [69]. Indeed, as shown in Fig. 10, the radical complexes are farther apart from each other, with the dsit ligand facing the Cp of a neighboring complex. Despite the large Se•••Se intermolecular distances (4.50–4.56 Å) when compared with twice the van der Waals radius of Se (≈ 2 Å), notable antiferromagnetic interactions settle within the chains of dyads, characterized by a uniform spin chain behavior ($J_{Mo}/k_B = -147$ K) in the Mo/dsit salt, an alternated chain behavior ($J_W/k_B = -113$ K, $\alpha = 0.5$) for the W/dsit complex. Note again the stronger antiferromagnetic interactions in the Mo salt, despite closely related structures and folding angles (θ) of the metallacycles ($MoSe_2C_2$, $\theta_{Mo} = 27.90(6)°$; WSe_2C_2, $\theta_W = 32.38(1)°$).

While the strong $TCNQF_4$ oxidant was needed to oxidize chemically the dmit (or dmid, dsit) complexes described above, TCNQ itself proved to be sufficient to generate the radical cations $[Cp_2Mo(dddt)]^+$ [70] and $[Cp_2W(dddt)]^+$ [38]. Salts

Table 1 Magnetic characteristics of the spin ladders in $[Cp_2Mo(dmid)][TCNQF_4]$ and $[Cp_2W(dmid)][TCNQF_4]$

	Δ[K]	$T(\chi_{max})$ [K]	$J_{\perp}/k_B$ [K]	$J_{//}/k_B$ [K]
$[Cp_2Mo(dmid)][TCNQF_4]$	74	70	−106.8	−40.7
$[Cp_2W(dmid)][TCNQF_4]$	13	17	−23.4	−15.6

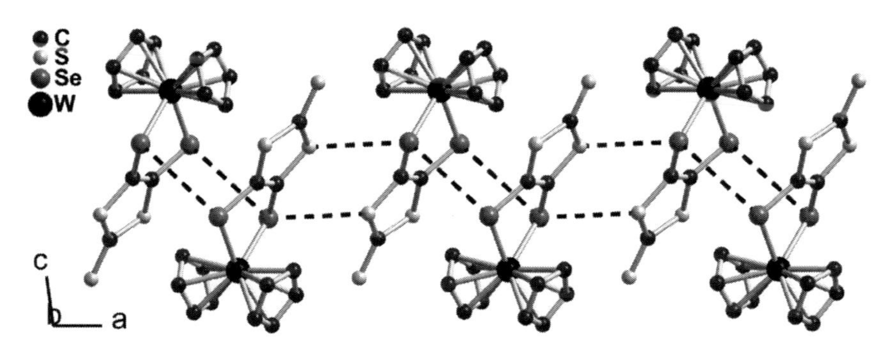

Fig. 10 Short S•••Se and Se•••Se contacts (*dotted lines*) within the chains of $[Cp_2W(dsit)]^+$ in $[Cp_2W(dsit)][TCNQF_4]$

were isolated with both complexes; they are isomorphous but small differences are observed, and particularly on the metallacycle folding angle which amounts to 32.1° in $[Cp_2Mo(dddt)]^+$, to 33.2° in $[Cp_2W(dddt)]^+$. As shown in Fig. 11, they adopt a structure organization reminiscent of that of $[(^tBuCp)_2W(dmit)][TCNQF_4]$ shown in Fig. 8 (left), with dyads of TCNQ radical anions, strongly associated into a face-to-face overlap (hence without any contribution anymore to the susceptibility). In between the $(TCNQ)_2^{2-}$ dyads, the radical cations stack along the *a* direction, forming alternated chains. Satisfactory fits of the temperature dependence of the magnetic susceptibility gave $J_{Mo}/k_B = -26$ K with $\alpha = 0.0-0.1$ for the Mo salt, $J_W/k_B = -16$ K with $\alpha = 0.5-0.6$ for the W salt. Since the two compounds are isomorphous, the differences between the two salts can probably be ascribed to different distributions of the spin density in the radical complexes, as already observed above in the dmid/dmit/dsit complexes.

This differential Mo/W behavior has been tentatively explained by an extended Hückel fragment analysis of the contribution of both Cp_2M and dithiolene fragment in the SOMO of the complexes. Indeed, as shown in Scheme 6, the LUMO of the Cp_2W fragment is closer in energy to the HOMO of the dithiolene fragment than the LUMO of the Cp_2Mo fragment. As a consequence, upon mixing, the HOMO of the Cp_2W(dithiolene) complexes have a stronger metal character than that of the Mo derivatives.

Finally, one other example of a spin ladder has been reported within these extensive series of $[Cp_2M(dt)]^+$ salts, in the AsF_6^- salt of $[Cp_2W(dsit)]^+$ [34]. We will see below that the other AsF_6^- (Sects. 3.3.1 and 3.3.2) salts of $[Cp_2Mo$

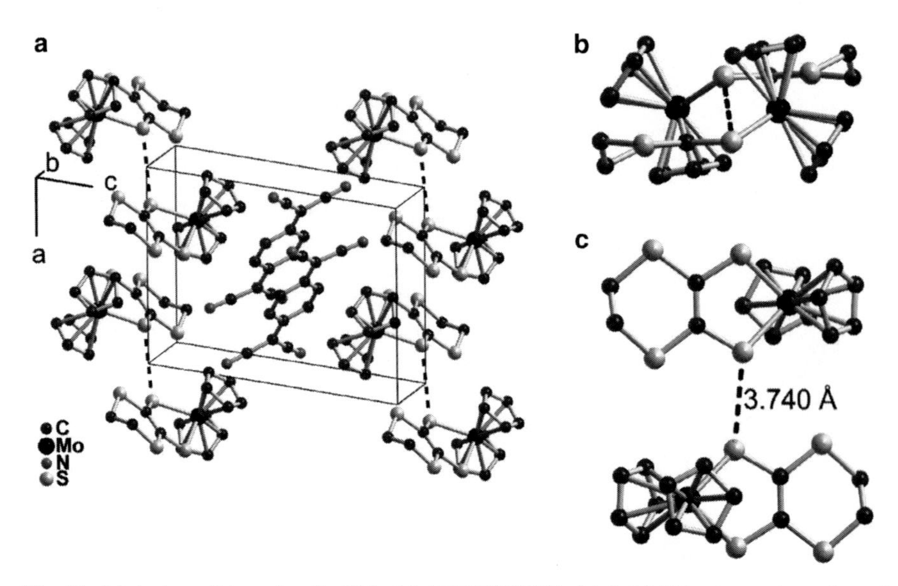

Fig. 11 (**a**) A view of the unit cell of [Cp_2Mo(dddt)][TCNQ]. (**b**) & (**c**) Side view and top view of the dyad association with the short intermolecular S•••S distance (*dotted line*) involving the S atoms of the metallacycle

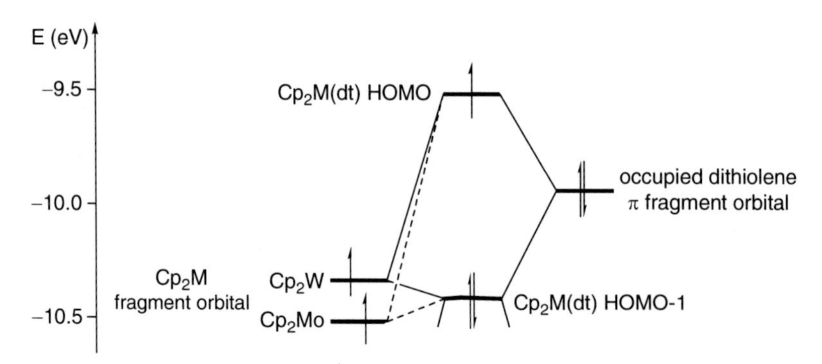

Scheme 6 Extended Hückel fragment analysis of the interaction between the partially occupied Cp_2M and occupied dithiolene fragment orbitals in d^1 [Cp_2M(dt)]$^{+\bullet}$ complexes, M = Mo, W (adapted from [69])

(dmit)], and [Cp_2W(dmit)] are isomorphous and crystallize without any folding of the metallacycle (θ = 0°) despite their d^1 character. Probably because of the presence of those heavy atoms such as W and Se, the [Cp_2W(dsit)][AsF_6] salt behaves differently and crystallizes with a strong folding [θ = 30.1(1)°]. The spin ladder motif found in its solid state structure is confirmed by the temperature dependence of the magnetic susceptibility.

3.2.2 The [CpNi(Dithiolene)] Complexes

The formally Ni^{III}, paramagnetic [CpNi(dithiolene)] complexes present a variety of intermolecular interaction networks, depending strongly on the geometry and steric requirements of the substituents on the dithiolate ligand. The absence of any counter ion, combined with a large spin density delocalization, leads systematically to the presence of relatively strong antiferromagnetic interactions and no compound among the 20 complexes explored so far exhibits a simple Curie-type law of noninteracting spins. The various complexes described below are collected in Scheme 7 together with their acronym.

A singlet-triplet behavior is found for those complexes such as [CpNi(oxdt)] [71]or [CpNi(F_2pdt)] [72] associated into dyads (Fig. 12), where the strong distortions from planarity of the dithiolate ligands hinder any other interdyad intermolecular contacts. The large Ni•••S and S•••S intermolecular distances lead to weak intermolecular interactions with J values of -29 or -8 cm^{-1} for [CpNi(oxdt)] and [CpNi(F_2pdt)] respectively.

On the other hand, [CpNi(dddt)] [59], [CpNi(mnt)] and [CpNi(tfd)] [73] give rise to the formation of spin chains but with very different overlap interaction motifs for the three compounds. Indeed, while [CpNi(dddt)] molecules interact side-by-side with neighboring molecules through lateral intermolecular S•••S contacts, [CpNi(mnt)] was found to stack with neighboring molecules into alternated chains with two different overlap interactions noted A and B in Fig. 13. Theoretical calculations (DFT) of the exchange coupling constants for the two interactions demonstrated unambiguously that the stronger one was interaction B with a short Ni•••S intermolecular contact, while interaction A within the face-to-face inversion centered motif was much smaller, due to a strong concentration of the spin density on the NiS_2 motif. As a consequence, short intermolecular distances within the A dyads do not correspond to any sizeable spin density overlap. [CpNi(tfd)] offers another new attractive feature as it forms a spin chain based on a Cp•••dithiolene overlap, as shown also in Fig. 13. DFT calculations of the exchange coupling constant $J_{calc} = -30$ cm^{-1} was in good agreement with that

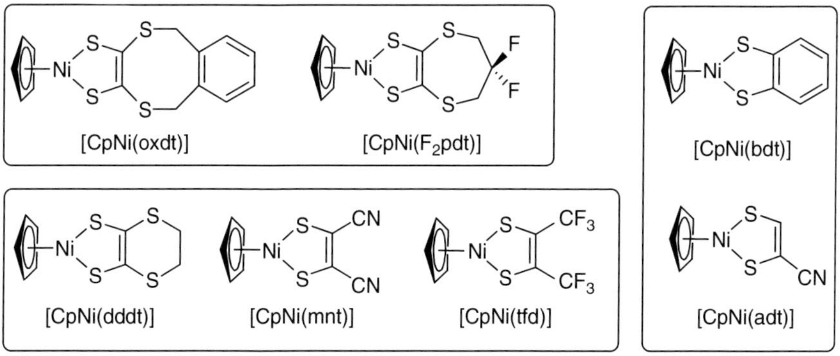

[CpNi(oxdt)] [CpNi(F_2pdt)] [CpNi(bdt)]

[CpNi(dddt)] [CpNi(mnt)] [CpNi(tfd)] [CpNi(adt)]

Scheme 7 Examples of [CpNi(dithiolene)] complexes

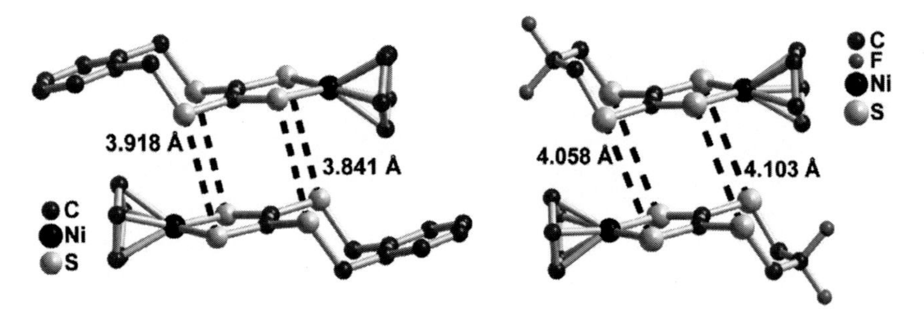

Fig. 12 View of the dyad association with shortest S•••S intermolecular distances for [CpNi(oxdt)] (*left*) and [CpNi(F_2pdt)] (*right*)

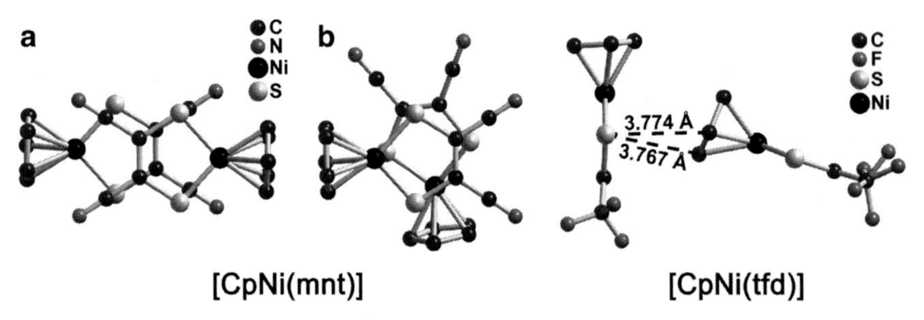

Fig. 13 Overlap interactions motif in [CpNi(mnt)] (**a**, **b**) and in [CpNi(tfd)]

determined experimentally from the fit of the temperature dependence of the magnetic susceptibility ($-43\ cm^{-1}$).

Finally, another type of intermolecular interaction was revealed in those complexes such as [CpNi(bdt)] [59] and [CpNi(adt)] [73]. The structure of those two compounds does not reveal any Ni•••S or S•••S interactions which could explain the strong antiferromagnetic interactions deduced from the temperature dependence of the spin susceptibility. However, they both exhibit a face-to-face interaction between the Cp rings (Fig. 14). DFT calculations of the exchange coupling constants confirmed that these Cp—Cp intermolecular interactions were indeed at the origin of the antiferromagnetic behavior, an original feature in organometallic chemistry, attributable to a sizeable delocalization of the spin density on the Cp ring (up to 20% in [CpNi(bdt)]).

3.3 Three-Dimensional Antiferromagnetic Ground State

We have seen above several examples where $[Cp_2M(dt)]^+$ (M = Mo, W) complexes organize in the solid state into *low dimensional* structures, leading to characteristic magnetic behaviors such as spin chains (eventually alternated) or spin ladders. The extensive use in later years of dithiolene ligands such as dmit or dddt was aimed at

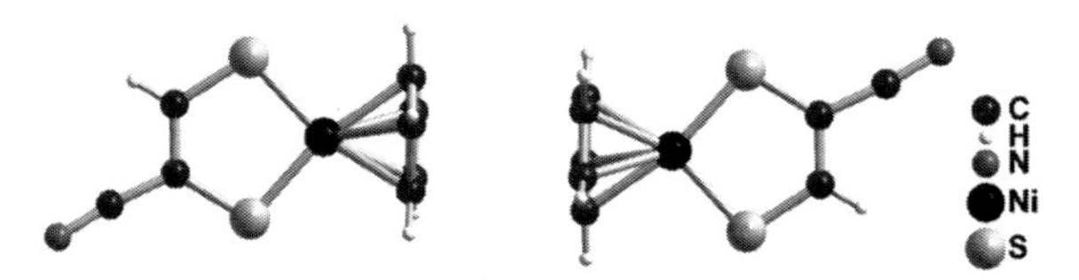

Fig. 14 Cp•••Cp overlap interaction in [CpNi(adt)]

favoring intermolecular interactions. In some cases, detailed below, this approach was so successful that a transition to an ordered three-dimensional antiferromagnetic ground state was observed, reflecting the presence of extended sets intermolecular interactions between the magnetic species.

3.3.1 The $[Cp_2M(dmit)]AsF_6$ Salts (M = Mo, W)

As mentioned above for $[Cp_2W(dsit)](AsF_6)$, it is possible to use electrocrystallization rather than chemical oxidation (with TCNQ or $TCNQF_4$) to crystallize the cation radical salts of the $[Cp_2M(dt)]^{+\bullet}$ species. This technique has been extensively used for the crystallization of mixed valence conducting systems derived from tetrathiafulvalene donor molecules and analogs [74]. It works particularly well when the oxidized material which precipitates at the electrode is itself a good conductor. The starting crystallites can then function as electrode themselves and the crystals grow into the solution. This technique, however, is not limited to conducting systems and, as exemplified below, has been successfully applied to 1:1 insulating phases. It consists in oxidizing the $Cp_2M(dt)$ complexes at the anode of a two-compartment cell, using a electrolyte soluble in organic solvents such as $^nBu_4NPF_6$, $^nBu_4NAsF_6$ or $^nBu_4NSbF_6$. A small constant current density allows for a regular crystal growth, and the crystals can be harvested on the anode after typically 1–2 weeks.

Under those conditions, the electrocrystallization of $[Cp_2Mo(dmit)]$ and $[Cp_2W(dmit)]$ in the presence of $^nBu_4NAsF_6$ afforded plate-like crystals of the 1:1 salts, formulated as $[Cp_2M(dmit)]AsF_6$, with M = Mo, W [32, 34]. These salts are isomorphous. They crystallize in the orthorhombic system, space group Cmcm, with the complex on a crystallographic site symmetry m2m (Fig. 15). As a consequence, the metallacycle is not folded at all ($\theta = 0°$), a peculiarity of those salts despite their formal d^1 character. This unfolded structure has, as a consequence, that the HOMO is essentially localized on the dithiolene moiety, as confirmed by the bond length evolution within the metallacycle, with a strong shortening of the C–S and concomitant lengthening of the C=C bonds, when compared with the neutral $Cp_2M(dmit)$ complexes or their oxidized but folded forms [22]. As shown in Fig. 15, each cation is surrounded by ten neighbors, without the formation of these dyads otherwise systematically encountered, as discussed above in Sect. 3.2.

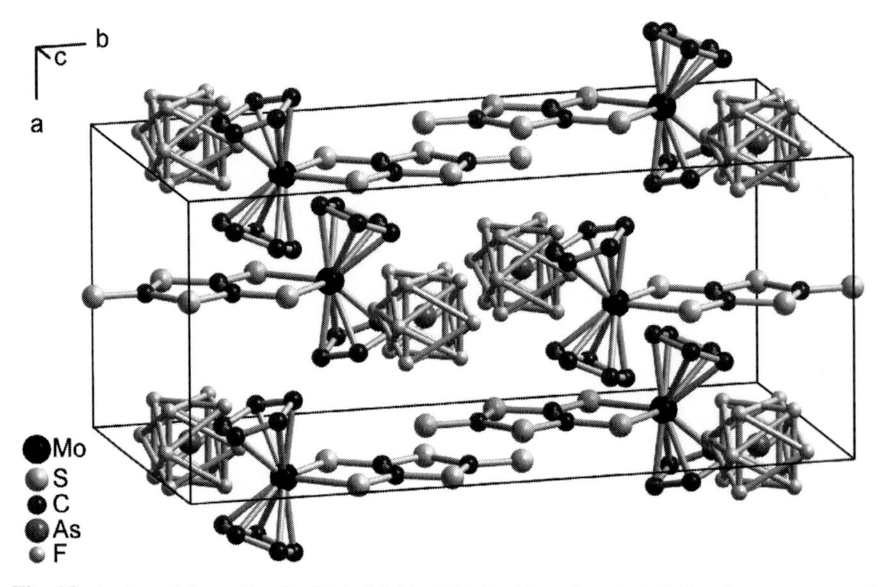

Fig. 15 A view of the unit cell of $[Cp_2Mo(dmit)]AsF_6$. Note that the AsF_6^- anions, represented here as polyhedra based on the F•••F distances, are disordered on a mirror plane

The temperature dependence of the magnetic susceptibility for these systems is characterized by a Curie-Weiss behavior with antiferromagnetic interactions, much stronger for the Mo than for the W analog: $\theta_{Mo}= -21$ K, $\theta_W= -3.4$ K. Upon cooling, a transition to an antiferromagnetic ground state is evidenced from the field dependence of the magnetic susceptibility below the Néel temperature T_N, with $T_N(Mo) = 9.5$ K and $T_N(W) = 3.5$ K. The observation of such an ordered 3D ground state confirms that, in these AsF_6^- salts, the radical cations interact with neighboring molecules through a three-dimensional set of interactions. In single crystal EPR experiments, one single line is observed with weak anisotropy in both salts which confirms that the spin density is preferentially localized on the dithiolene. On the other hand, the room temperature line width is much larger and highly anisotropic ($\Delta H_a = 270$ G, $\Delta H_b = 400$ G, $\Delta H_c = 325$ G) for the W salt than for the Mo salt ($\Delta H_a = 20$ G, $\Delta H_b = 22$ G, $\Delta H_c = 30$ G). This Mo/W isomorphous substitution has other strong effects on the magnetic properties of the salts.

In the paramagnetic regime, the evolution of the EPR line width and g value show the presence of two transitions, observed at 142 and 61 K in the Mo salt, and at 222 and 46 K in the W salt. Based on detailed X-ray diffraction experiments performed on the Mo salt, the high temperature transition has been attributed to a structural second-order phase transition to a triclinic unit cell with apparition of a superstructure with a modulation vector $q_1 = (0, 1/2, 1/2)$. Because of a twinning of the crystals at this transition, it has not been possible to determine the microscopic features of the transition, which is probably associated to an ordering of the anions, which are disordered at room temperature, an original feature for such centrosymmetric anions. This superstructure remains present down to the Néel temperature, indicating that the second transition is not related to a structural phase transition.

Similarly, in the antiferromagnetic ground state, the two Mo and W complexes differ not only by the value of the Néel temperature but also but by the valuc of the spin-flop field. It amounts indeed to 6,500 G in the Mo salt, but to 19,000 G in the W salt, a sign of the stronger spin-orbit coupling in the latter. Indeed, the spin-flop field is directly related to the magnetic anisotropy in the AF state and has two main origins in such molecular systems: the spin orbit coupling and the dipolar interactions. Since very similar dipolar interactions are expected for two salts which are isomorphous, the larger spin-flop field of the W complex reflects indeed a larger spin orbit coupling contribution in $[Cp_2W(dmit)]AsF_6$.

3.3.2 The $[Cp_2Mo(dmit)]X$ Salts ($X = PF_6^-$, AsF_6^-, SbF_6^-) and Their Solid Solutions

We have described above the evolution of the magnetic properties of the $[Cp_2M(dmit)]AsF_6$ salts upon isomorphous Mo/W substitution. Another possibility offered by this attractive series is the isomorphous substitution of the counter ion, that is PF_6^- vs AsF_6^- vs SbF_6^-. Electrocrystallization experiments conducted with $[Cp_2Mo(dmit)]$ and the three different electrolytes afforded an isomorphous series, with a smooth evolution of the unit cell parameters with the anion size [32]. This cell expansion with the anion size leads to decreased intermolecular interactions between the $[Cp_2Mo(dmit)]^{+\bullet}$ radical cation, as clearly seen in Table 2 from the decreased Curie-Weiss temperatures and Néel temperatures (associated with the transition they all exhibit to an AF ground state).

Despite their isomorphous character, a notable difference appeared between the PF_6^- salt on one hand and the AsF_6^- and SbF_6^- salt on the other. Indeed, as already mentioned above, the AsF_6^- and SF_6^- salts exhibit upon cooling two second-order phases transitions, at 142 and 61 K in the former, at 175 and 40 K in the latter, before the transition at T_N to the AF ground state. On the other hand, the PF_6^- salt exhibits a second-order phase transition at 120 K, followed now by a first-order transition at 89 K. The latter was unambiguously identified from hysteresis loops in the temperature dependence of the EPR line width, and also associated by an abrupt change in the g tensor values. While the high temperature transition is associated for the three anions to a $q_1 = (0, 1/2, 1/2)$ superlattice in the triclinic space group, the first-order transition of the PF_6^- salt was associated with the appearance of a novel

Table 2 Evolution of the structural and magnetic characteristics of the $[Cp_2Mo(dmit)]X$ salts ($X = PF_6^-$, AsF_6^-, SbF_6^-) with the anion size (B_{SF} stands for the spin-flop field in AF state)

	Unit cell volume (Å^3)	θ (K)	T_N (K)	B_{SF} (G)
$[Cp_2Mo(dmit)]$ PF_6	1863.5(3)	−37	11.5	7,000
$[Cp_2Mo(dmit)]$ AsF_6	1895.5(3)	−21	9.5	6,500
$[Cp_2Mo(dmit)]$ SbF_6	1959.0(3)	−14	7.5	6,200

superlattice at a wave vector $q_2 = (0, 1/2, 1/2)$. The main question which arose then was: why is there such a different behavior between the PF_6^- salt and the two other AsF_6^- and SbF_6^- salts? In order to get some answers, solid solutions associating PF_6^- and AsF_6^- on one hand and PF_6^- and SbF_6^- on the other were prepared by the same electrocrystallization technique using appropriate relative concentrations of the anions in the electrolyte [75]. A schematic phase diagram obtained for both solid solutions $[Cp_2Mo(dmit)](AsF_6)_x(PF_6)_{1-x}$ and $[Cp_2Mo(dmit)](SbF_6)_x(PF_6)_{1-x}$ is given in Fig. 16.

Without going into details, we observe the high temperature phase (Phase I), two related phases (II and IV) with the same (0, 1/2, 1/2) superlattice q_1, phase III with the $q_2 = (0, 1/2, 0)$ superlattice. The first-order line ends with liquid/gas like critical point at x_C. This diagram presents a clear analogy with the phase diagram of a pure compound. The analogs of phases II and IV would be respectively the gas and the liquid, separated from each other by the first order transition line up to the critical point C. Phase III would be the analog of the solid phase. This behavior has been analyzed in the frame of the Landau theory with a compressible model involving two modes of distortion [75].

3.3.3 The Cp*M(Dithiolene)$_2^{\bullet}$ Series (M = Mo, W)

The oxidation by electrocrystallization of the d^2 anionic complexes $[Cp^*M(dmit)_2]^-$ with M = Mo or W afforded in both cases the neutral radical species $[Cp^*M(dmit)_2]^{\bullet}$ on the electrode [42, 43] as a crystalline material. The two compounds are isomorphous. As shown in Fig. 17, the absence of any counter ion

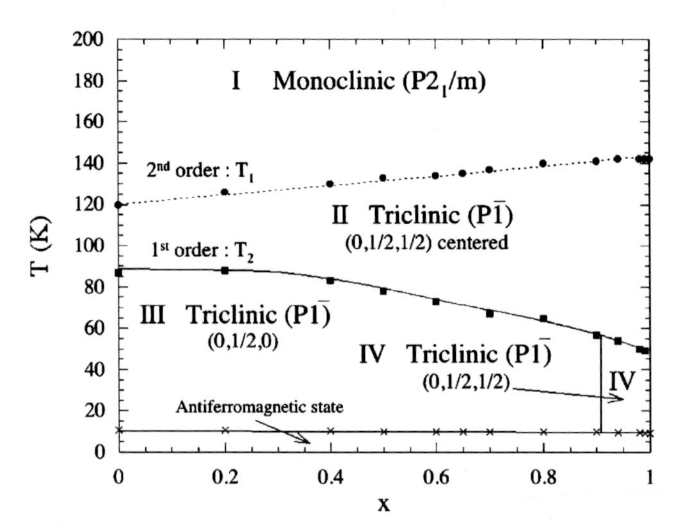

Fig. 16 Schematic phase diagram obtained from the experimental studies of the two solid solution series $[Cp_2Mo(dmit)](AsF_6)_x(PF_6)_{1-x}$ and $[Cp_2Mo(dmit)](SbF_6)_x(PF_6)_{1-x}$ (from [75], reproduced with the kind permission of EDP Sciences)

allows for the setting of dmit/dmit and Cp*/dmit interactions in the solid state, affording a truly three-dimensional interaction network.

In the 50–300 K temperature range, the magnetic susceptibility is well fitted with a Curie-Weiss law with Curie-Weiss temperatures of −38 and −20 K for the Mo and W complexes respectively, demonstrating the presence of strong antiferromagnetic interactions, with the same trends as observed above between $[Cp_2Mo(dmit)]^{+\bullet}$ and $[Cp_2W(dmit)]^{+\bullet}$ salts, that is, systematically weaker interactions with the tungsten complexes. DFT calculations of the spin density distribution in these complexes (Table 3) confirm the above observations, with a spin density on the metal which goes from almost zero on the Mo atom to 15% on the W atom, while it remains constant and very weak (≈4%) on the Cp* ligand. As a consequence, the spin density on the two dmit ligands is notably decreased in the W complex, a possible origin of the weaker antiferromagnetic interactions.

A three-dimensional set of intermolecular interactions is further confirmed by the observation of a transition to an antiferromagnetic ground state in both radical complexes, at a Néel temperatures of 8 (Mo) and 4.5 K (W), in accordance with the difference of Curie-Weiss temperatures between both complexes. Note also the spin-flop field in the antiferromagnetic state, found at 5.5 kG in $[Cp^*M(dmit)_2]^\bullet$ and at 8 kG in $[Cp^*W(dmit)_2]^\bullet$, a consequence of the stronger spin orbit coupling in the latter.

3.3.4 The CpNi(Dithiolene) Series

Analogous to the d^1 $Cp^*M(dmit)_2^\bullet$ series of neutral paramagnetic complexes described above, the formally d^7 $CpNi(dithiolene)^\bullet$ complexes provide another

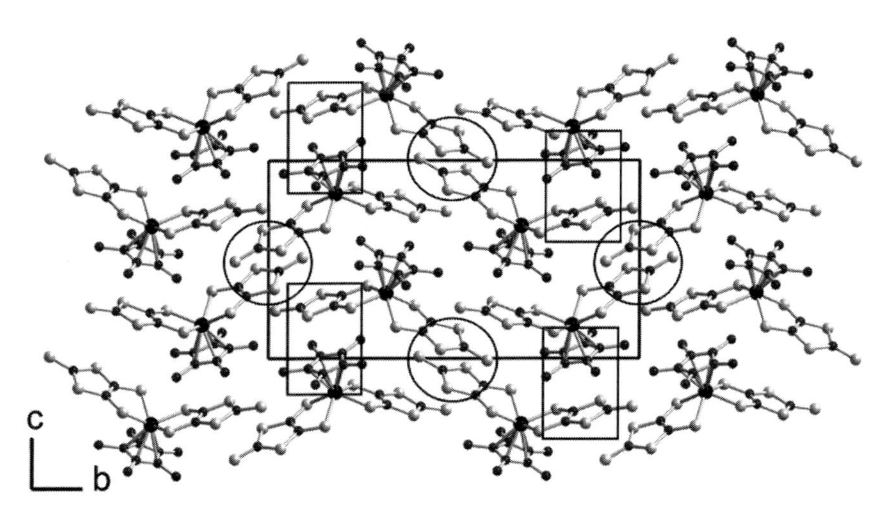

Fig. 17 A view of the interaction network on $[Cp^*W(dmit)_2]$ showing the dmit•••dmit (*circles*) and the Cp•••dmit (*squares*) interactions

Table 3 Spin density distribution on the metal atom and summed over all atoms of each of the three ligands in $[Cp^*M(dmit)_2]^•$, M = Mo, W [43]

	M	dmit	dmit	Cp*
$[Cp^*Mo(dmit)_2]^•$	0.001	0.471	0.489	0.037
$[Cp^*W(dmit)_2]^•$	0.150	0.403	0.407	0.040

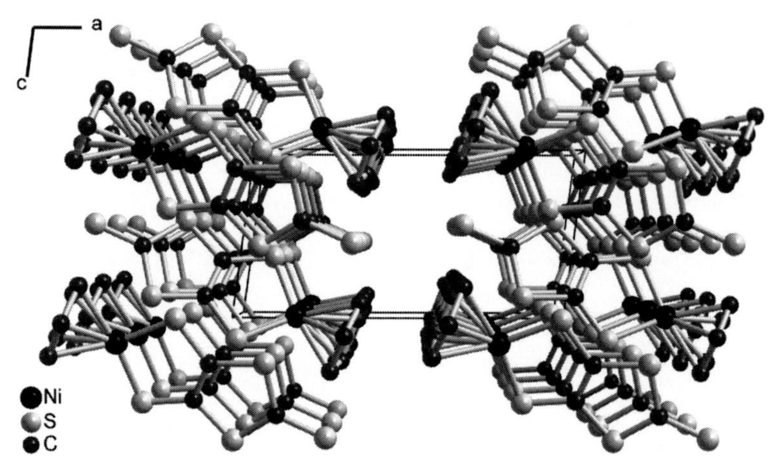

Fig. 18 A view of the unit cell of CpNi(dmit) showing the layered structure with additional Cp•••S interlayer interactions

family of neutral complexes where the absence of counter ion should favor a three-dimensional set of intermolecular interactions and an antiferromagnetic order as ground state. This is indeed the case in CpNi(dmit) and CpNi(dsit) which are isomorphous and crystallize in the $P2_1$ space group [57, 76]. As shown in Fig. 18, the complexes are organized into layers, and are without any face-to-face overlap of the planar dithiolene moieties.

An extended set of intermolecular interactions thus develops within these layers with short S•••S and S•••Se contacts, complemented by weaker interlayer interactions corresponding to Cp•••S interactions. The temperature dependence of the magnetic susceptibility shows strong antiferromagnetic interactions with Curie-Weiss temperatures of −66 and −33 K for the dmit and dsit complexes respectively. The susceptibility of both compounds exhibits a maximum at 43 K (dmit) and 23 K (dsit), and below a transition to an antiferromagnetic ground state with $T_{\text{Néel}}$(dmit) = 27 K and $T_{\text{Néel}}$(dsit) = 18 K, as demonstrated from the field dependence of the magnetic susceptibility below $T_{\text{Néel}}$ in both compounds. This AF ground state with elevated Néel temperatures (for such molecular systems) confirms the presence of a three-dimensional set of intermolecular interactions, favored by the absence of any counter ions. Other series of CpNi(dithiolene) complexes are currently being investigated to evaluate the origin and generality of such strong antiferromagnetic interactions.

4 Summary and Outlook

We have shown here that different series of heteroleptic $[Cp_nM(dt)_m]$ complexes are stable and isolable in the solid state in their paramagnetic state. The different geometries these radical molecules adopt in the solid state are intimately related to the extent of delocalization of the spin density between the metal centre, the Cp and dithiolene ligands. Of particular note is the highly variable folding of the MS_2C_2 metallacycle encountered in most complexes. The magnetic behavior of these complexes in the crystalline state reflects this variability of the spin density distribution, from strongly localized interactions into diamagnetic dyads, to extended magnetic structures as one-dimensional spin chains and ladders, or three-dimensional ordered antiferromagnets. Most remarkably, and in contrast to the "classical" square-planar bis(dithiolene) complexes, the intermolecular interactions between these radical, nonplanar complexes are not limited here to S•••S interactions but Cp•••S and Cp•••Cp contacts were also shown to be particularly efficient in some cases for mediating antiferromagnetic interactions, an original and promising feature within the broad field of molecular magnetism.

Acknowledgments Our own contribution to this work has only been possible through an intense and long lasting collaboration with Dr R. Clérac and Prof C. Coulon from the Centre de Recherches Paul Pascal (CRPP-CNRS) in Bordeaux (France). Their strong involvement and exceptional knowledge of molecular magnetism were much appreciated. We also thank Drs T. Cauchy and M. Nomura for their invaluable contribution on the [CpNi(dt)] complexes and specially Dr Cauchy for careful reading of the manuscript.

References

1. Wudl F, Wobshall D, Hufnagel EJ (1972) J Am Chem Soc 94:670
2. Ferraris J, Cowan DO, Walatka V, Perlstein JH (1973) J Am Chem Soc 95:948
3. Jérome D, Mazaud A, Ribault M, Bechgaard K (1980) J Phys Lett 41:L95
4. Yagubskii EB, Schegolev IF, Laukhin VN, Kononovich PA, Kartsovnik MV, Zvrykina AV, Buravov LI (1984) JEPT Lett 39:12
5. Margadonna S, Prassides K (2002) J Solid State Commun 168:639
6. For a complete review, see the thematic issue of Chem Rev, Batail P (ed) (2004) Chem Rev 104(11)
7. Robertson N, Cronin L (2002) Coord Chem Rev 227:93
8. Kato R (2004) Chem Rev 104:5319
9. Mueller-Westerhoff UT, Vance B (1987) Comprehensive coordination chemistry, Chap. 16–5. Pergamon, Oxford
10. Kirk ML, McNaughton RL, Helton ME (2004) In: Stiefel EI (ed) Dithiolene chemistry. Prog Inorg Chem 52:111
11. Lim BS, Fomitchev DV, Holm RH (2001) Inorg Chem 40:4257

12. Faulmann C, Cassoux P (2004) In: Stiefel EI (ed) Dithiolene chemistry. Prog Inorg Chem 52:399
13. See, for example, Beswick CL, Schulman JM, Stiefel EI (2004) In: Stiefel EI (ed) Dithiolene chemistry. Prog Inorg Chem 52:55
14. Roger M, Arliguie T, Thuéry P, Fourmigué M, Ephritikhine M (2005) Inorg Chem 44:584
15. Roger M, Arliguie T, Thuéry P, Fourmigué M, Ephritikhine M (2005) Inorg Chem 44:594
16. Livage C, Fourmigué M, Batail P, Canadell E, Coulon C (1993) Bull Soc Chim Fr 130:761
17. Cummings SD, Eisenberg R (2004) In: Stiefel EI (ed) Dithiolene chemistry. Prog Inorg Chem 52:315
18. Burgmayer SJN (2004) In: Stiefel EI (ed) Dithiolene chemistry. Prog Inorg Chem 52:491
19. McMaster J, Tunney JM, Garner CD (2004) In: Stiefel EI (ed) Dithiolene chemistry. Prog Inorg Chem 52:539
20. King RB (1963) J Am Chem Soc 85:1587
21. Fourmigué M (1998) Coord Chem Rev 178/180:823
22. Fourmigué M (2004) Acc Chem Res 37:179
23. Harris HA, Kanis DR, Dahl LF (1991) J Am Chem Soc 113:8602
24. Guyon F, Fourmigué M, Audebert P, Amaudrut J (1995) Inorg Chim Acta 239:117
25. Lauher JW, Hoffmann R (1976) J Am Chem Soc 98:1729
26. Guyon F, Lenoir C, Fourmigué M, Larsen J, Amaudrut J (1994) Bull Soc Chim Fr 131:217
27. Cranswick MA, Dawson A, Cooney JJA, Gruhn NE, Lichtenberger DL, Enemark JH (2007) Inorg Chem 46:10639
28. Guyon F, Fourmigué M, Clérac R, Amaudrut J (1996) J Chem Soc Dalton Trans 4093
29. Viard B, Amaudrut J, Sala-Pala J, Fakhr A, Mugnier Y, Moise C (1985) J Organomet Chem 292:403
30. Goldberg DP, Michel SLJ, White AJP, Williams DJ, Barrett AGM, Hoffman BM (1998) Inorg Chem 37:2100
31. Michel SLJ, Goldberg DP, Stern C, Barrett AGM, Hoffman BM (2001) J Am Chem Soc 123:4741
32. Clérac R, Fourmigué M, Gaultier J, Barrans Y, Albouy PA, Coulon, C (1999) Eur Phys J B 9:431
33. Clérac R, Fourmigué M, Coulon C (2001) J Solid State Chem 159:413
34. Domercq B, Coulon C, Fourmigué M (2001) Inorg Chem 40:371
35. Arliguie T, Fourmigué M, Ephritikhine M (2000) Organometallics 19:109
36. Roger M, Belkhiri L, Thuéry P, Arliguie T, Fourmigué M, Boucekkine A, Ephritikhine M (2005) Organometallics 24:4940
37. Jourdain IV, Fourmigué M, Guyon F, Amaudrut J (1999) Organometallics 18:1834
38. Jourdain IV, Fourmigué M, Guyon F, Amaudrut J (1998) J Chem Soc Dalton Trans 483
39. Herberhold M, Jin GX, Milius W (1994) Z Anorg All Chem 620:1295
40. Takacs J, Kiprof P, Herrmann WA (1990) Polyhedron 9:2211
41. King RB, Bisnette MB (1967) Inorg Chem 6:469
42. Fourmigué M, Coulon C (1994) Adv Mater 6:948
43. Domercq B, Coulon C, Feneyrou P, Dentan V, Robin P, Fourmigué M (2002) Adv Funct Mater 12:359
44. Nomura M, Sasaki S, Fujita-Takayama C, Hoshino Y, Kajitani M (2006) J Organomet Chem 691:3274
45. Adams H, Gardner HC, McRoy RA, Morris MJ, Motley JC, Torker S (2006) Inorg Chem 45:10967
46. Cleland WE Jr, Barnhart KM, Yamanouchi K, Collison D, Mabbs Fe, Ortega RB, Enemark JH (1987) Inorg Chem 26:1017
47. Belkhiri L, Arliguie T, Thuéry P, Fourmigué M, Boucekkine A, Ephritikhine M (2006) Organometallics 25:2782
48. Arliguie T, Fourmigué M, Ephritikhine M (2000) Organometallics 19:109
49. Stephan DW (1992) Inorg Chem 31:4218

50. Fourmigué M, Perrocheau V (1997) J Mater Chem 7:2235
51. Guyon F, Lucas D, Jourdain IV, Fourmigué M, Mugnier Y, Cattey H (2001) Organometallics 20:2421
52. Nomura M, Fourmigué M (2008) Inorg Chem 47:1301
53. Guyon F, Jourdain IV, Knorr M, Lucas D, Monzon T, Mugnier Y, Avarvari N, Fourmigué M (2002) Eur J Inorg Chem 2026
54. Mori H, Nakano M, Tamura H, Matsubayashi G, Mori W (1998) Chem Lett 729
55. Mori H, Nakano M, Tamura H, Matsubayashi G (1999) J Organomet Chem 574:77
56. Baird HW, White BM (1966) J Am Chem Soc 88:4744
57. Faulmann C, Delpech F, Malfant I, Cassoux P (1996) J Chem Soc Dalton Trans 2261
58. Nomura M, Okuyama R, Fujita-Tayakama C, Kajitani M (2005) Organometallics 24:5110
59. Nomura M, Cauchy T, Geoffroy M, Adkine P, Fourmigué M (2006) Inorg Chem 45:8194
60. Grosshans P, Adkine P, Sidorenkova H, Nomura M, Fourmigué M, Geoffroy M (2008) J Phys Chem A 112:4067
61. Guyon F, Amaudrut J, Mercier MF, Shimizu K (1994) J Organomet Chem 465:187
62. Domercq B, Fourmigué M (2001) Eur J Inorg Chem 1625
63. Bleaney B, Bower KD (1952) Proc R Soc Lond Ser A214:451
64. Faulmann C, Cassoux P (2004) In: Stiefel EI (ed) Dithiolene chemistry. Prog Inorg Chem 52:399
65. Bonner JC, Fisher ME (1964) Phys Rev A 135:640
66. Kahn O (1993) Molecular magnetism, Chap. 11. VCH, New York
67. Estes WE, Gavel DP, Hatfield WE (1978) Inorg Chem 17:1415
68. Hatfield WE (1981) J Appl Phys 52:1985
69. Fourmigué M, Domercq B, Jourdain IV, Molinié P, Guyon F, Amaudrut J (1998) Chem Eur J 4:1714
70. Fourmigué M, Lenoir C, Coulon C, Guyon F, Amaudrut J (1995) Inorg Chem 34:4979
71. Nomura M, Fourmigué M (2007) New J Chem 31:528
72. Nomura M, Geoffroy M, Adkine P, Fourmigué M (2006) Eur J Inorg Chem 5012
73. Cauchy T, Ruiz E, Jeannin O, Nomura M, Fourmigué M (2007) Chem Eur J 13:8858
74. Batail P, Boubekeur K, Fourmigué M, Gabriel JCP (1998) Chem Mater 10:3005
75. Clérac R, Fourmigué M, Gaultier J, Barrans Y, Albouy PA, Coulon C (1999) Eur Phys J B 9:445
76. Fourmigué M, Avarvari N (2005) Dalton Trans 1365

Index

G

H

I

L

M

N

O

P

R

S

T

U

V